Fundamentals and Applications of Crop and Climate Science

f Ahmad
ion of Agrometeorology, Faculty of
culture
e-Kashmir University of Agricultural
ces and Technology of Kashmir
gar, Jammu and Kashmir, India

Biswas
ol of Environmental Sciences
ersity of Guelph
ph, ON, Canada

Gazi Mohammad Shoaib Shah
Faculty of Horticulture
Sher-e-Kashmir University of Agricultural
Sciences and Technology of Kashmir
Srinagar, Jammu and Kashmir, India

N 978-3-031-61461-3   ISBN 978-3-031-61459-0   (eBook)
://doi.org/10.1007/978-3-031-61459-0

Latief Ahmad • Gazi Mohammad Sh[...]
Asim Biswas

Fundamentals and Applications of Crop [and] Climate Science

Springer

Preface

The *Fundamentals and Applications of Crop and Climate Science* has been written exclusively for students of Agriculture and its allied disciplines. In this book, we embark on a journey to unravel the intricate relationship between crops and climate, exploring its fundamental principles and practical applications in agricultural contexts.

Our fascination with the interplay between crops and climate stems from a deep appreciation for the complexity and resilience of natural systems. Through our research and experiences, we have come to understand the profound impacts of climate variability and change on crop growth, development, and productivity, as well as the critical role that crops play in shaping and responding to their environment.

In the pages that follow, we delve into the mechanisms by which crops perceive and adapt to climatic stimuli, from temperature fluctuations and water availability to atmospheric composition. We explore how these interactions influence crop physiology, yield potential, and environmental sustainability, providing insights that are essential for informing agricultural management and decision-making.

As authors, we hope that this book serves as a valuable resource for researchers, students, policymakers, and practitioners alike, empowering them to navigate the complexities of crop-climate dynamics and harness their potential for positive change under new age agricultural system in the world. Together, let us embrace the challenges and opportunities presented by the nexus of crops and climate, working toward a more resilient, sustainable, and food-secure future for generations to come.

Srinagar, Jammu and Kashmir, India — Latief Ahmad
Srinagar, Jammu and Kashmir, India — Gazi Mohammad Shoiab Shah
Guelph, ON, Canada — Asim Biswas

Contents

Chapter 1
Introduction to Agronomy

Abstract Agronomy is the science and practice of crop production and field management. It is an amalgamation of many sciences and views the agriculture from an integrated and holistic perspective. It combines the knowledge and technology from the fields of botany, ecology, chemistry, genetics, pest management, water management, weed science, and soil science in order to improve and manage the major food crops of the world. It studies various aspects of crop production like soil properties and how they affect the crop growth, which nutrients are needed by crops, how to control the pests, diseases, and weeds. This chapter introduces you to agronomy and gives details of the history, development, and scope of agriculture. Various agroclimatic zones of India and concept of tillage are also discussed.

Keywords Agronomy · Agriculture · Tillage · Agroclimatic Zone · HYV · Topography · Tilth

1.1 History, Development, and Scope of Agriculture

Agriculture as we all know comes from two Latin words, *"**Ager**"* and *"**Colo**"* meaning *"**field**"* and *"**cultivate**,"* respectively. So the term agriculture means the cultivation of land or field. It is the science and art of cultivating crops and raising livestock.

1.1.1 History and Development

History of agriculture dates back to thousands of years when there was a strict gender-based demarcation of the day-to-day chores. Men were the ones doing the hunting and gathering, while women used to stay indoors and participate in chores like feeding and taking care of young ones. It went like that for many thousand years until a woman discovered that plants grow from seeds. Around 11,500 years ago, people started gradually settling down in one place and practice growing cereal and

L. Ahmad et al., *Fundamentals and Applications of Crop and Climate Science*,
https://doi.org/10.1007/978-3-031-61459-0_1

tuber crops. By 2000 years ago, majority of the population on earth was dependent on agriculture. This led to domestication of plants and wild animals. The first crop to be domesticated was rice, and the first animal is believed to be a dog.

The earliest civilizations based on agriculture were settled around the rivers Tigris, Euphrates (Mesopotamia), and Nile (Egypt). In 5500 BC, the settlements of Mesopotamia developed simple irrigation systems. By 6000 BC, new variety of rice was developed in Egypt and South Asia.

The periods of major developments in agriculture started from the eighteenth century with the invention of horse-drawn seed drill, by Jethrow Tull. The eighteenth and nineteenth century saw development of new varieties of field crops along with the invention of many machines like cotton gin, mechanical reaper, and thresher. In this period, agriculture started getting shape of a subject that needed to be taught. This led to establishment of various research institutes, universities, and extension centers.

In recent years, agriculture has turned into a business opportunity rather than a source of survival. Genetic, hydrological, mechanical, and business developments in agriculture have ensured upliftment, sustainability, and equal opportunities at life for a farmer.

1.1.2 Scope of Agriculture

Agriculture is a vast, dynamic, and ever-changing subject. The advent of new mechanical, genetic, and chemical technologies for enhancing crop production and minimization of crop losses makes the scope of agriculture diverse and immense throughout the world especially in the agrarian-driven economies. Following are the points that illustrate the scope of agriculture:

- Agriculture is the most important enterprise in the world. Without agriculture, the life on earth won't exist.
- Agriculture is indispensable from the existence of humans and animals.
- The population expansion is at its peak in the twenty-first century, but the land under cultivation is more or less the same; hence, to feed the ever-growing population, new high-yielding varieties need to be developed along with the sustainable use of preexisting ones.
- Along with the use of HYVs, the focus is being shifted to intensive agriculture and to make the most out of the given piece of land sustainably.
- With the increased cost of energy, there has been a major shift of attention toward no-tillage practices.
- As the unemployment rates have been soaring post-pandemic, the agriculture sector still manages to employ more people than any other sector of the world.
- With respect to India, agriculture employs 58% of the countries workforce and contributes to 17% to the GDP. Agriculture sector ensures not only the food security of the citizens of India but also the national security and integrity.

1.2 Agroclimatic Zones of India

The Planning Commission in 1989 (presently NITI Aayog) has divided the country into 15 agroclimatic regions based on soil type, topography, agroclimatic features, temperature, and rainfall variation. These zones are further divided into 72 homogeneous subzones.

The 15 agroclimatic zones are as follows:

- Zone 1: Western Himalayan Region: Jammu and Kashmir, Uttar Pradesh
- Zone 2: Eastern Himalayan Region: Assam, Sikkim, West Bengal and all North-Eastern states
- Zone 3: Lower Gangetic Plains Region: West Bengal
- Zone 4: Middle Gangetic Plains Region: Uttar Pradesh, Bihar
- Zone 5: Upper Gangetic Plains Region: Uttar Pradesh
- Zone 6: Trans-Gangetic Plains Region: Punjab, Haryana, Delhi, and Rajasthan
- Zone 7: Eastern Plateau and Hills Region: Maharashtra, Uttar Pradesh, Orissa, and West Bengal
- Zone 8: Central Plateau and Hills Region: MP, Rajasthan, Uttar Pradesh
- Zone 9: Western Plateau and Hills Region: Maharashtra, Madhya Pradesh, and Rajasthan
- Zone 10: Southern Plateau and Hills Region: Andhra Pradesh, Karnataka, Tamil Nadu
- Zone 11: East Coast Plains and Hills Region: Orissa, Andhra Pradesh, Tamil Nadu, and Pondicherry
- Zone 12: West Coast Plains and Ghat Region: Tamil Nadu, Kerala, Goa, Karnataka, Maharashtra
- Zone 13: Gujarat Plains and Hills Region: Gujarat
- Zone 14: Western Dry Region: Rajasthan
- Zone 15: The Islands Region: Andaman and Nicobar, Lakshadweep (Fig. 1.1)

Zone 1: Western Himalayan Region
- It includes the states of Jammu and Kashmir, Himachal Pradesh, and Uttarakhand's Kumaun-Garhwal region. Showing great variation in relief, this area has an average annual rainfall of 150 cm.
- Because of the high rainfall and snow-covered mountain peaks of the Ganga, Yamuna, Jhelum, Chenab, Satluj, and Beas, the region has perennial streams. They provide irrigation water to canals as well as low-cost hydropower to agriculture and industry.
- The summers are mild, but winter season experiences severe cold conditions. The average July temperature is 5–30 °C, while that in January, it can be as low as 0 °C to −4 °C.
- The important crops of this region are wheat, maize, potato, and barley. Apples and pears are the major temperate fruits grown in parts of Kashmir and Himachal Pradesh.

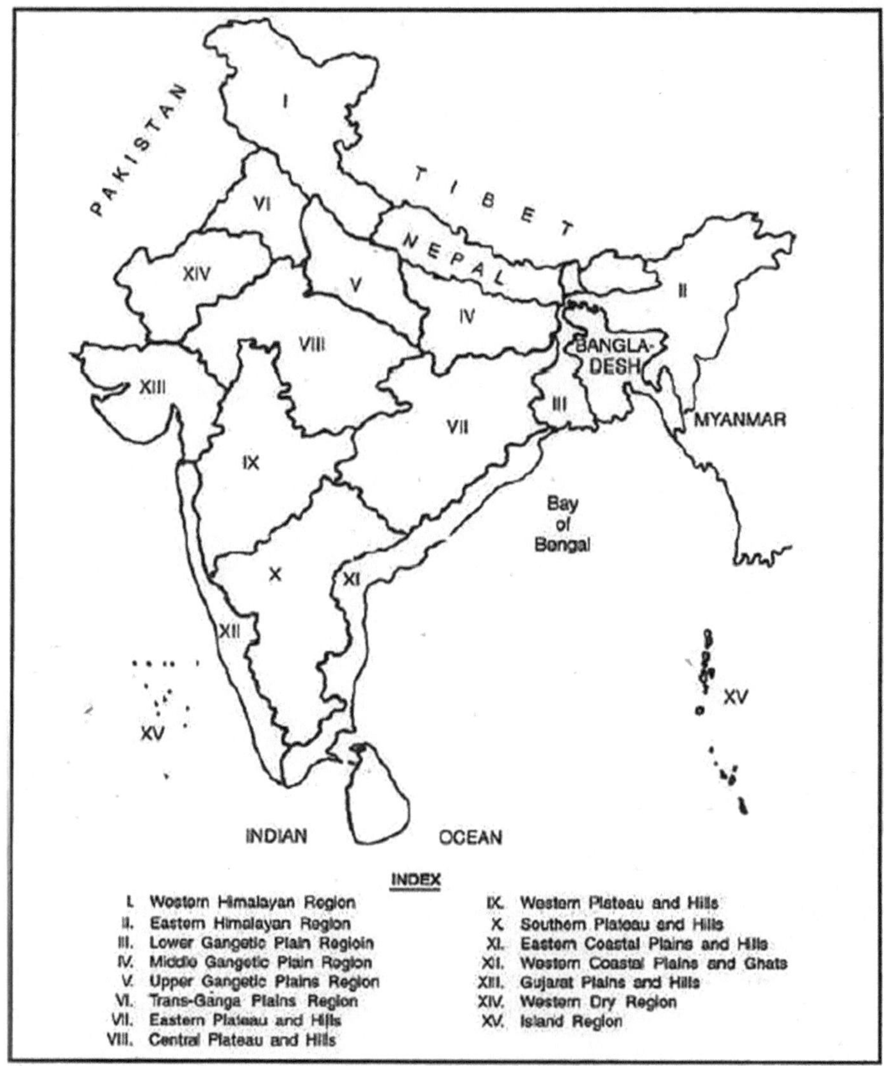

Fig. 1.1   Source: Fundamentals of Agronomy. (Dr. H H Patel)

Zone 2: Eastern Himalayan Region

- Sikkim, the Darjeeling area (West Bengal), Arunachal Pradesh, the Assam hills, Nagaland, Meghalaya, Manipur, Mizoram, and Tripura comprise the Eastern Himalayan region.
- They are is distinguished by its rugged topography, dense forest cover, and sub-humid climate.

- The soil in this area is brownish, layered, and not very fertile. Jhum also known as shifting cultivation is practiced in nearly one-third of the cultivated area, and food crops are grown primarily for sustenance.
- The rainfall exceeds 200 cm; temperatures range from 25 to 33 °C in July and 11 to 24 °C in January.
- Rice, potato, maize, tea, and fruits including orange, pineapple, lime, and litchi are the main crops.

Zone 3: Lower Gangetic Plains Region
- This region encompasses eastern Bihar, West Bengal, and the Assam Valley. The average annual rainfall in this area ranges between 100 and 200 cm. The temperature ranges from 26 to 41 °C in July and 9 to 240 °C in January.
- With a high water table, the region has adequate groundwater storage. Irrigation is primarily provided by wells and canals.
- Rice is the main crop, which can yield three crops in a year. Other important crops include jute, maize, potato, and pulses.

Zone 4: Middle Gangetic Plains Region
- It incorporates parts of Uttar Pradesh (eastern) and Bihar (western).
- Rice, maize, millets, wheat, barley, mustard, and potato are the important crops of the region.
- Lying in the Ganga basin, it is a fertile alluvial plain.
- The average temperature ranges from 26 to 41 °C in July and 9 to 24 °C in January. It receives an annual rainfall of 100–200 cm.

Zone 5: Upper Gangetic Plains Region
- Central and Western parts of Uttar Pradesh plus Hardwar and Udham Nagar districts of Uttarakhand form a part of this region.
- The region is characterized by sandy loam soil and practice of traditional agriculture method.
- An important feature of this region is the subhumid continental type of climate. The temperature ranges from 26 to 41 °C in July and 7–23 °C in January. It receives rainfall of 75–150 cm annually.
- Major crops are wheat, rice, sugarcane, millets, maize, gram, barley, oilseeds, pulses, and cotton.

Zone 6: Trans-Gangetic Plains Region
- This zone is spanned across Haryana, Punjab, Chandigarh, Delhi, and Rajasthan's Ganganagar district.
- The region is also known by Sutlej-Yamuna plains.
- The chief source of irrigation is private tube wells and canals.
- It is characterized by semiarid type of climate with average temperature in July ranging from 26 to 42 °C and in January temperature ranging from 7 to 22 °C. The average annual rainfall ranges from 70 to 125 cm.
- Important crops are wheat, sugarcane, cotton, gram, millets, pulses, and oilseeds.

Zone 7: Eastern Plateau and Hills Region
- It includes Chotanagpur plateau, the Rajmahal Hills, the Chhattisgarh plains, and Dandakaranya.
- The primary source of irrigation is rainwater since there are no perennial rivers here.
- The region is characterized by red and yellow soils.
- The region experiences temperatures ranging from 26 to 34 °C in July and 10 to 27 °C in January and annual rainfall ranging from 80 to 150 cm.
- Rice, millets, maize, oilseeds, ragi, gram, and potato are grown.

Zone 8: Central Plateau and Hills Region
- It encompasses Bundelkhand, Baghelkhand, Bhander Plateau, Malwa, and Vindhyachal.
- Red, yellow, and black soils are found in this region. There is scarcity of water resources.
- Semiarid climate in western parts and subhumid climate in eastern part. Annual rainfall ranges from 50 100 cm.
- Millets, wheat, gram, oilseeds, cotton, and sunflowers are mainly grown in these regions.

Zone 9: Western Plateau and Hills Region
- It encompasses southern part of the Malwa plateau and Deccan plateau.
- The major crops are grown in rainfed fields, and only 12.4% of the area is irrigated.
- Temperature between 24–41 °C is experienced in July and 6–23 °C in January and average annual rainfall of 25–75 cm.
- Oranges, bananas, and grapes are popular in this region. Other important crops are jowar, cotton, sugarcane, rice, bajra, wheat, gram, pulses, potato, groundnut, and oilseeds.

Zone 10: Southern Plateau and Hills Region
- This region includes the interior Deccan, parts of southern Maharashtra and Karnataka, Andhra Pradesh, and Tamil Nadu.
- Considerable amount of agriculture is practiced here. Climate in this region is semiarid type. Only 50% of the area is under cultivation.
- Major crops: Millets, oilseeds, pulses, coffee, tea, cardamom, and spices.
- Temperature ranges from 26 to 42 °C in July and between 13 and 21 °C in January. Average rainfall is 50–100 cm.

Zone 11: Eastern Coastal Plains and Hills
- The important coastal plains of India, Coromandel, and Sircar coasts make up this zone.

- It is characterized by subhumid maritime and receives an annual rainfall of 75–150 cm.
- The soils of the region are facing the problem of alkalinity.
- Rice, jute, tobacco, sugarcane, maize, millets, groundnut, and oilseeds are the major crops grown in this region, and it accounts for about 20.33% of rice and 17.05% of groundnut production of the country.

Zone 12: West Coast Plains and Ghats Region
- Malabar and Konkan coasts and the Sahyadris make up this region. This region is best for prawn culture.
- Coconut, oilseeds, sugarcane, millets, and cotton are important crops. Its famous for plantation crops and spices.
- This is a humid area with annual rainfall exceeding 200 cm and average temperatures ranging from 26 to 32 °C in July and 19 to 28 °C in January.

Zone 13: Gujarat Plains and Hills Region
- Kathiawar and fertile valleys of the Mahi and Sabarmati rivers encompass this zone.
- Marine fishing and dry land farming are suited for this area.
- Groundnut, oilseeds, rice, millets, and cotton are major crops. This region is mainly famous for producing oilseeds.
- The region is arid and semiarid.

Zone 14: Western Dry Region
- This zone includes the area west of Aravallis.
- Famine and drought are frequent occurrences.
- The temperature ranges from 28 to 45 °C in June and 5 to 22 °C in January. The area is characterized by scanty vegetation and absence of perennial rivers.
- The main crops grown here are bajra, jawar, moth, and gram.
- Because of the varying temperatures, this region is ideal for livestock.

Zone 15: The Islands Region
- The region consists of Andaman and Nicobar and Lakshadweep Islands. Equatorial climates are typical of this area.
- Along the coasts, sandy soils are found, while valleys and lower slopes have clayey loam.
- The average temperature in July is 30 °C and 25 °C in January.
- Major crops are rice, areca nut, turmeric, and cassava.

1.3   Brief Classification of Crops

(a) Classification on the Basis of Use

• **Edible crops:**

> (1) Cereal crops : wheat, bajra
>   (monocot)
> (2) Pulses crops (dicot): gram
> (3) Fruits crops : mango, banana
> (4) Vegetable crops: brinjal, potato
> (5) Legumes : groundnut
> (6) Oil seeds: sesamum, groundnut

• **Special Purpose crops:**

> In agriculture field, there are various
> crops that are grown with a special
> purpose under certain circumstances.
> They are sometimes grown for
> consumption and sometimes for other
> purposes.
> (1)Green manure crops: dhaincha
> (2)Silage crops: oats
> (3)Cover crops: grasses
> (4)Trap crops: marigold
> (5)Companion crops: berseem and
> barley

• **Non-Edible crops:**

> Those crops that are grown to produce
> goods for manufacturing like fiber for
> clothing and in affiliated agriculture
> industries, rather than for food
> consumption.
> (1)Fiber crops: jute, hemp
> (2)Non edible oilseed crops: karanja,
> castor
> (3)Forage crops: soundal, fodder maize
> (4)Narcotic crops: cannabis, tobacco
> (5)Colour crops: saffron

(b) Classification Based on Duration of Crops

Annual crops
The term annual refers to the crops that complete their life cycle in one year. These crops, germinate, grow, flower, come to bearing and die within a year, for example, sorghum, rice, maize and brinjal

Biannual crops
Those crops that complete their life within 2 years but more than 1 year. They develop a root system and leaves in first year and flower in the second year. Hence, biennials complete their lifecycle in two years, for example, banana, papaya, sugarcane

Perennial crops
Those crops that complete their life in more than 2 years, for example, mango, sapota, guava

(c) **Classification based on the season:**

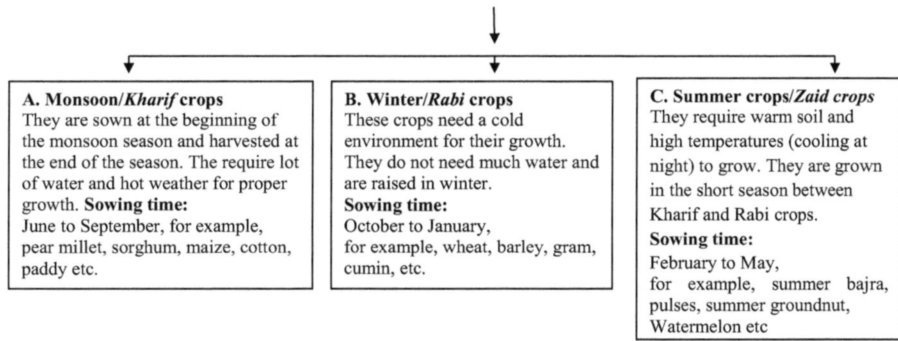

1.4   Concept of Tillage: Tilth and Factors Affecting It

Tillage is an age old concept. It being an indispensable part of agriculture is as old as agriculture itself. Jethro Tull is considered the father of tillage and propounded a theory that soil needs to be broken down into fine particles for the plants to absorb it, although his theory was proved wrong later on (Fig. 1.2).

Seeds, in order to germinate, need loose, friable, and a well-drained soil with physical factors. After harvesting the crop, the soil becomes clumpy, hard, and compact and is full of previous crop stubble, weeds, and other unwanted entities. In order to make the soil suitable for the sowing and subsequent germination of next generation of crop seed, tillage needs to be done.

Fig. 1.2 Jethro Tull. (nzdl.org)

Tillage is the physical manipulation of the soil done to break the soil clods into finer masses and to make the soil suitable for seed germination. On the other hand, tilth comes off as a result of tillage.

1.4.1 Objectives of Tillage

Some objectives of tillage are mentioned below:

(a) Tillage helps in breaking the compact soil layer leading to better root permeability.
(b) It helps to incorporate green manure crops, weeds, organic manures, and fertilizers, into the soil.
(c) Tillage helps in destruction of harmful soil organisms, like dormant fungal spores, pest pupae, and microbes, by exposing them to the sun.
(d) Tillage also improves soil aeration, which results in better organic matter decomposition leading to higher nutrient availability for plants. Along with this, better soil aeration also helps in quick degradation of pesticides and herbicides residues in the soil.
(e) One of the most important uses of tillage is to incorporate noxious weeds like (*Chenopodium album*) into the soil, ultimately decomposition.
(f) Tillage provides an excellent seed bed for the sowing and germination of the new crop seed.
(g) It helps in providing better seed—soil contact allowing optimum water flow to seedling.

1.4.2 Factors Affecting Tilth

These are the major factors affecting the tilth of the soil:

(a) Soil structure

It is one of the most important factors affecting the soil tilth. The most desirable one is called "crumb," in which the soil aggregates are loose, water stable, and optimally porous.
(b) Organic matter

Soils with high organic matter content if provided with optimum moisture form the desirable crumb structure.
(c) Weather

Weather is the one of the major determining factors of the soil structure.
(d) Soil texture

Although soil texture cannot be altered, but its influence can directly be seen on soil porosity, friability, and permeability, all of which are correlated with tilth.

Horse Drawn   **Seed Drill**
Source:

inventors.findthebest.com

Horse drawn Seed drill. (Source: inventors.findthe**best.com**)

MCQs

1. The origin of the world agriculture is:

   - Latin
   - Greek
   - Roman
   - Persian

2. How many agroclimatic zones are there in India?

   - 6
   - 16
   - 9
   - 15

3. ______ and ______ are the major temperate fruits grown in parts of Kashmir and Himachal Pradesh.

   - Louquat and peach
   - Peach and nectarines
   - Apple and pear
   - Cherry and grapes

4. Western Bihar comes under which agroclimatic region?

   - Middle Gangetic Plains Region
   - Trans-Gangetic Plains Region
   - Eastern Plateau and Hills Region
   - Western Plateau and Hills Region

5. Which one of the below mentioned crop is used as a companion crop?

  - Wheat
  - Jute
  - Rice
  - Berseem

6. Who is considered the father of tillage?

  - Jethro Tull
  - Norman Borlaugh
  - M. S. Swaminathan
  - None of the above

7. Wheat is a ______ crop

  - Zaid
  - Kharif
  - Rabi
  - None of the above

8. Which of the following crops is used as a green manure crop

  - Wheat
  - Dhaincha
  - Rice
  - Barley

9. Sowing time of Zaid crop is

  - June to November
  - February to May
  - October to January
  - September to December

10. How much does agriculture contribute to Indian GDP?

  - 11%
  - 7%
  - 17%
  - 29%

References

Chandrasekaran, C., Annadurai, K., & Somasundaram, E. (2010). *A textbook of agronomy* (pp. 1–3). New Age International Pvt. Ltd..
http://ecoursesonline.iasri.res.in/mod/page/view.php?id=1550
http://jalshakti-dowr.gov.in/agro-climatic-zones
Patel, H. H. (2020). *Fundamentals of agronomy* (pp. 63–66). Scientific Publishers.

Chapter 2
Soils

Abstract Soil is the living, biologically active, porous, and nonrenewable dynamic resource providing ecosystem services critical for life. This chapter will introduce readers to soil and various concepts pertaining to soil like soil profile, genesis of soil, characteristics of soil, and soil pollution. Air movement, water movement, temperature, and pH are correlated with soil. Specific section has been dedicated to soils of India. This chapter provides a general understanding of various soil concepts along with the problems that soil undergoes, either naturally or because of humans. It provides an overall insight of soil, for better land management decisions.

Keywords Genesis · Soil profile · Pedology · Edaphology · Porosity · Problematic Soils · Reclamation

2.1 Introduction to Soils: Edaphological and Pedological Concepts

Soil is the uppermost part of the earth's surface, which for the most part is loose and falls to pieces when applied pressure on. It is made up of small particles of varying size and shape. It is mainly formed because of humus formation, withering of rocks, and addition and removal of various materials over a prolonged period of time. Soils vary from place to place in terms of production potential, nutrient content, humus content, appearance, water holding capacity, etc.

Pedology: It encompasses the origin, description, and classification of soil. It involves the study of soil as an integral part of ecosystem, without much focus on its production capacity.

Edaphology: The study of soil with a focus on plant production is edaphology. It includes the properties of soil, which affect the organisms, especially the plants.

L. Ahmad et al., *Fundamentals and Applications of Crop and Climate Science*,
https://doi.org/10.1007/978-3-031-61459-0_2

2.2   Soils of India

- Alluvial soil
- Red soil
- Black soil
- Laterite soil
- Arid soil
- Forest soil
- Mountainous soil (Fig. 2.1)

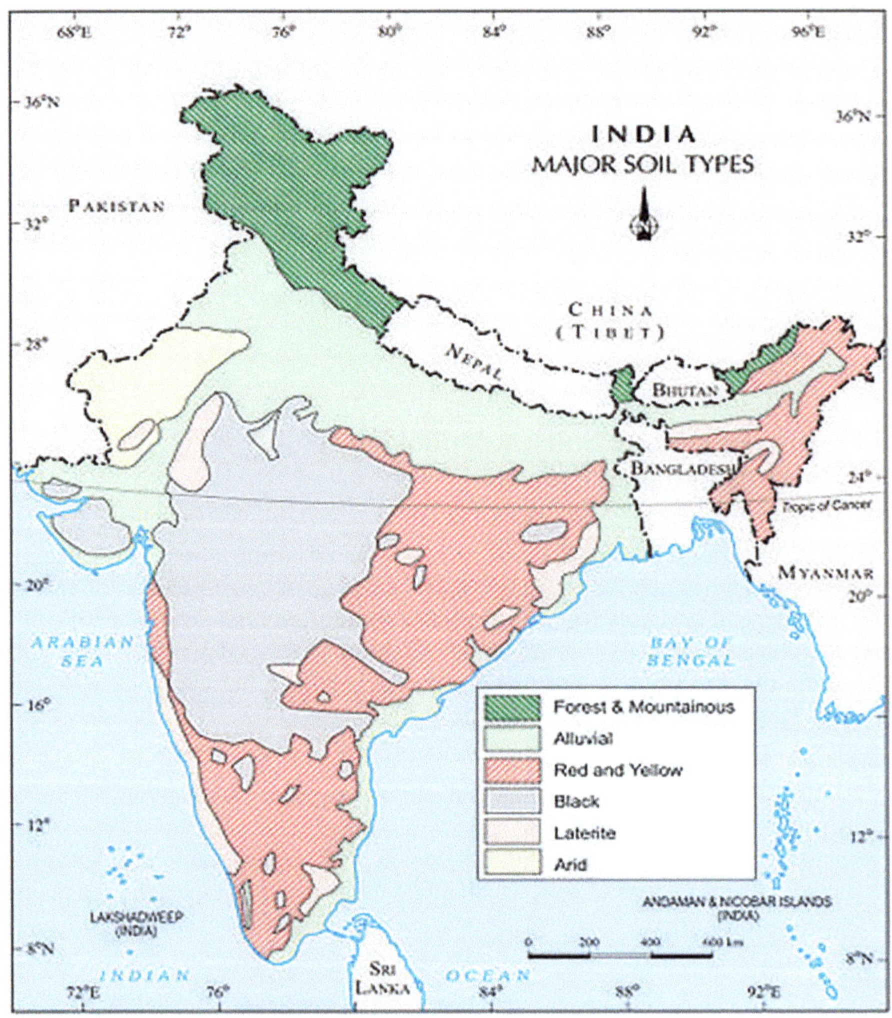

Fig. 2.1   Source: commons.wikimedia.org

1. *Alluvial Soil*

It is the most widely available soil in India, covering up to 43% of the land. It is highly fertile, humus, and organic matter-rich and is widespread in the North-Indian Plains and river valleys. Indus-Ganga-Brahmaputra plain and Narmada-Tapi plain are some examples. It is rich in potash and poor in phosphorus. New alluvial is called Khadar, and old alluvial is called Bhangar. Its texture is sandy to silty loam or clay. It is light gray to ash gray in color.

2. *Red Soil*

It covers 18.5% of the Indian landmass. It has a porous and friable structure. It is mainly seen in low rainfall areas. It is devoid of lime, phosphate, nitrogen, and potash. Its texture is sandy to clay and loamy. It is red in color because of the presence of ferrous oxide.

3. *Black Soil*

It is best suited for cotton cultivation, hence also termed as Regur soil. Most of our Deccan plateau is covered by black soil. It is also called mature soil. It has high WRC. It is rich in iron, lime, calcium, and potassium. It is deficient in nitrogen and phosphorus. It is clayey in nature with a deep black to light black color.

4. *Laterite Soil*

Its name has a Latin origin meaning "black." It is soft when wet and hard when dried. It formed when excessive leaching takes place in the areas with high temperature and high rainfall. It is rich in iron and aluminum and deficit in nitrogen, potassium, potash, and lime. It is red in color.

5. *Arid Soil*

It is also called desert soil. It is mainly deposited by fast-blowing winds. It has high salt content. It is present in desert areas where there is no or little rainfall. It has high content of kankar or calcium carbonate. It is highly sandy in texture, and its color varies from red to brown.

6. *Forest Soil*

It is acidic in nature with low humus content. It is present in areas that receive high rainfall.

7. *Mountain Soil*

It is also called immature soil and is present in mountainous regions. It is highly acidic with low humus content.

2.3   Soil Genesis and Soil Profile

2.3.1   *Soil Genesis*

Soil genesis is also termed as "pedogenesis," which is derived from a Greek word "**Pedon**" and "**genesis**" meaning "**soil**" and "**birth**." It is the process of soil formation as an effect of topography, environment, and history.

Soil formation includes the following processes:

1. Translocation
2. Organic changes
3. Podzolization
4. Gleying
5. Desilication

1. *Translocation*

   Translocation includes various types of systematic physical movements, which are predominantly in the downward direction. Various processes included in translocation are as follows:

   (a) *Leaching*

   It occurs in the areas where precipitation exceeds evaporation. It is the downward movement of water, in wet regions, toward the bedrock at the bottom. It takes the minerals that are present in the dissolved form away from the surface.

   (b) *Eluviation*

   It is the downward percolation of water leading to washing down of soluble materials to the bottom regions. It leaves behind a deprived horizon.

   (c) *Illuviation*

   It is the complete opposite of the eluviation. The process called illuviation refers to the deposition of nutrients and other dissolved substances, removed from the upper layers, through eluviation, into deeper horizons.

   (d) *Calcification*

   The formation and accumulation of calcium carbonate in warm environments is called calcification. It enriches the B-horizon of soil.

   (e) *Salinization*

   It occurs in the areas where evaporation exceeds precipitation. In this process, water moves up the soil and takes up the minerals and deposits it in the upper surfaces.

2. *Organic Changes*

   These changes are seen on the surface and follow a definite pattern. The organic substances are broken down into simpler ones by fungi, algae, protists, and worms leading to degradation, which in turn causes humification, leaving behind dark amorphous humus.

This is followed by mineralization. The further decomposition makes the humus release organic compounds into the soil.

3. *Podzolisation*

The degradation of clay minerals in the strong acid soil is called podzolization. It results in binding of iron complexes with the organic substances of the soil. They accumulate in the dark subsurface layers of the soil.

4. *Glaying*

When anaerobic conditions prevail, iron complexes are either reduced or completely removed from the soil. It is more observed in marshy soils.

5. *Desilication*

As the name indicated, it includes the removal of silica because of powerful weathering and leaching of nutrients. It leads to reduction in the organic content of the soil.

2.3.2   Soil Profile

Soil profile is defined as the vertical section of the soil, depicting various layers at the bottom of which the unaffected bedrock or parent material is found.

These layers are called horizons, and the main horizons in the soil are O, A, B, C, and R. And the major ones are only A, B, and C.

The study of soil profile is important for soil scientists because it tells a lot about the soils present in the history, origin of soil under investigation, source of the soil, etc. (Fig. 2.2).

The first look at a soil profile reveals five master horizons, O, A, B, C, and R, but they, in some cases, reveal further divisions within themselves in the form of sub-horizons.

O-Horizon It is not common in all soils, and it is mostly present in forest soils and absent in arable soils, desert soils, and grasslands. The word "O" in "O-Horizon" stands for organic as this layer consist of decaying and decomposed organic matter.

It has two sub-horizons:

O_1: In this part, decaying plant and animal residues can be recognized from the naked eye.

O_2: In this region, decaying organic matter cannot be recognized from the naked eye.

A-Horizon It is the topsoil as its present on top. It is a mineral horizon, and it witnesses the accumulation of vast amounts of organic matter and minerals like clay, iron, and aluminum from the parent material. It proves an optimum environment to sustain for organisms like fungi, bacteria, nematodes, and worms. It frequently suffers *additions* and *losses*.

Fig. 2.2 Soil profile.
(Modified from Pidwirny,
2006)

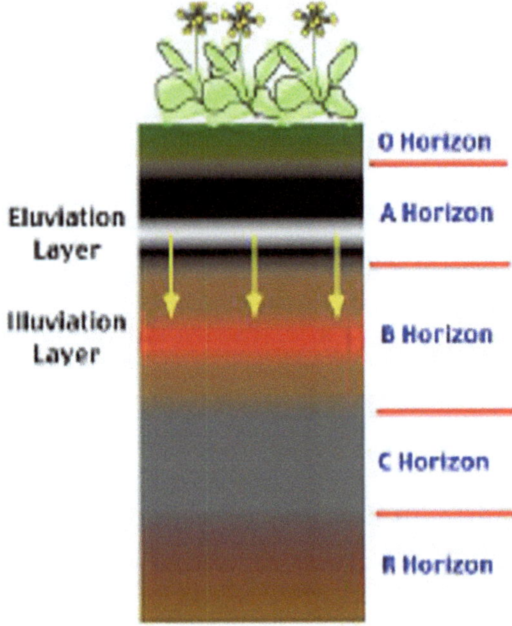

It has three sub-horizons:

A_1: It is the topmost layer, and it sits right next to the surface. It is rich in humified organic matter as compared to lower levels and hence comes off a bit darker in color.

A_2: It is the layer where maximum eluviations of organic matter, clay, iron, and aluminum oxides are observed. Losses of these pave way for accumulation of sand, silt, and quartz. It is generally lighter in color when compared to horizons above and below it.

A_3: It is the transitional layer between the A master horizon and B master horizon. It resembles more to the layers above it, in terms of physical properties, then to the layers below it.

B-Horizon It is a zone of accumulation called illuviation. It is also called subsoil. Minerals like clay and sesquioxide are deposited in this horizon, imparting a dark red color to it, which looks very distinct visually from the layers lying above and below it, making the B-Horizon easily identifiable. It frequently suffers *transformation* and *additions*.

It also has three sub-horizons like that of A-Horizon:

B_1: It is a transitional layer in between B and A horizon. It resembles more to A than B in terms of structure and properties.

B₂: It witnesses higher organic matter deposition and hence gives off rich, dark color as compared to its counterpart in the above horizon. The minerals like clay and aluminum oxides leach down the soil and get deposited here.

B₃: It is a transitional layer between B and C. It is more like B_2 than C.

C-Horizon This horizon contains parent material. It remains relatively unaffected by the soil forming processes. It is present outside the zone of major biological activity. In some cases, deposits of magnesium, calcium, and sulfates have been observed. There are no dominant *addition* and *losses* seen in this horizon, although minimal mineral processes like oxidation and reduction of iron are reported.

R-Horizon It contains the underlying bedrock and may or may not differ from the parent material of the C-Horizon. Almost no biological activity is observed in this region.

2.4 Soil Physics (A)

Soil physics includes all the physical properties of soil, like texture, structure, color, etc. These physical properties affect various chemical and biological processes taking place in the soil. In this section, we are going to discuss the major physical properties of soil.

2.4.1 Soil Texture

Soil texture refers to the proportion of three soil separates, namely, clay, silt, and sand, in the soil.

These soil separates are formed because of the actions of physical and chemical weathering on rocks and minerals over a prolonged period of time leading to the formation of particles like gravel, sand, stones, silt, and clay (Foth, 1991). Soil texture governs the processes like the, penetration of roots, water holding capacity, etc. Along with this, soil texture also determines the fertility of the soil and workability in the fields. One must always remember that the soil texture cannot be easily altered and is considered as the basic property of soil.

The three soils separates, sand, silt, and clay, are classified according to their size by various societies and organizations throughout the world. The most important and widely accepted ones are of United States Department of Agriculture System and International Soil Science Society System. They are illustrated in Table 2.1:

According to the percentage of soil constituents (silt, sand, and clay) present, the soil has been divided into 12 textural classes, represented by a soil textural triangle given by USDA Natural Resources Conservation Service (Fig. 2.3).

Table 2.1 Classification of soil separates as per their size

Separate	Diameter, mm[a]	Diameter, mm[b]	Number of particles per gram	Surface area in 1 gram, cm^2
Very coarse sand	2.00–1.00		90	11
Coarse sand	1.00–0.50	2.00–0.20	720	23
Medium sand	0.50–0.25		5,700	45
Fine sand	0.25–0.10	0.20–0.02	46,000	91
Very fine sand	0.10–0.05		722,000	227
Silt	0.05–0.002	0.02–0.002	5,776,000	454
Clay	Below 0.002	Below 0.002	90,260,853,000	8,000,000

[a]United States Department of Agriculture System
[b]International Soil Science Society System
The surface area of platy-shaped montmorillonite clay particles determined by the glycol retention method by Sor and Kemper. (See Soil Science Society of America Proceedings, Vol. 23, p. 106, 1959.) The number of particles per gram and surface area of silt and the other separates are based on the assumption that particles are spheres and the largest particle size permissible for the separate

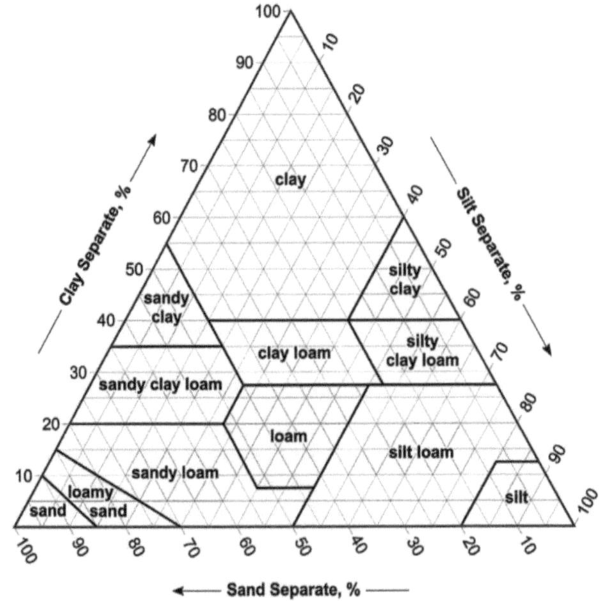

Fig. 2.3 Soil textural triangle. (USDA Natural Resources Conservation Service)

2.4.2 *Soil Structure*

As mentioned above, the soil texture is indicative of the size of the soil aggregates. The soil structure on the other hand determines the arrangement and organization of soil particles. Soil structure governs the processes like water movement, soil aeration, porosity, soil temperature, etc. Soil structure can somewhat also influence the extent of penetration of roots into the soil (Foth, 1991). Moreover, soil structure has a direct effect on the microbial activities in the soil as the microbes exist as a function of soil air, temperature, and water availability.

It can more or less be altered with different practices like ploughing, pulverizing, cultivating, and manuring, unlike soil texture that cannot be changed by any practice or procedure.

The primary particles of soil are usually grouped together in the form of aggregates. Naturally formed aggregates are called *pegs*, while as the man-made mass of soil is called a *clod*.

There are majorly four types of soil structures found on Earth:

- *Platy*: In this type, the peds are arranged in a horizontal manner just like leaflets or plates. The dimensions of horizontal axis are larger than that of vertical ones. Soil compaction can lead to platy structure (Mehra, 2004); this type can inhibit root penetration to a great extent. It is inherited from the parent material.
- *Prismatic*: In this type, the vertical axis shows relatively more prominence than Platy. When the tops are rounded, the structure is termed as columnar when the tops are flat/plane, level, and clear-cut prismatic.
- *Blocky*: All the three dimensions are of the same size. It is more or less six faced in a three dimensional system. When the faces are flat and the edges are with sharp angularity, the structure is named as angular blocky. When the faces and edges are mainly rounded, it is termed subangular blocky.
- *Spheroidal*: The peds are round, loose, and separated from each other. It is divided into crumbs and granular type. When the granules are porous, it is called crumbs type, and when nonporous in nature, it is termed as granular type (Fig. 2.4).

Fig. 2.4 Types of soil structure. (Mehra, 2004)

2.4.3 Soil Porosity

Soil porosity refers to the percentage of that part of soil that is not occupied by organic matter or soil aggregates itself. There are spaces present in the soil, called "soil pores." The area that air and water occupy in a soil is called the pore space. The arrangement of soil particles like sand, silt, and clay determines the amount of pore space in a soil.

Large particle size and larger pore spaces are characteristics of sandy soils (macropores). Although clay soils have many times more pore spaces than sandy soils, they nonetheless contain much smaller particles and much smaller pore spaces (micropores). Factoring in the organic matter, the higher its presence, the more pore space there is.

Macropores have a diameter of 75–5000 μm, while micropores have a diameter of 5–30 μm (Clothier et al., 2008).

Due to gravity and mass flow, air and water flow easily within macropores. On the other hand, air and water travels very gradually through micropores and only through capillary movement and diffusion.

Sandy soils have more macropores, while clayey soils reveal more micropores. Thus, while having a higher pore space, sandy soils have quick water and air movement caused by macropores, whereas clayey soils have slower air and water movement caused by micropores, despite having a higher total pore space. Fifty percent porosity and an equal amount of macropores and micropores are characteristics of loamy soils.

2.4.4 Soil Density

Soil density can be studied under two broad forms, that is, *bulk density* and *particle density*. Bulk density, which includes both the particles and the pore space, is the density of the bulk soil in its natural form, as opposed to particle density, which is the average density of the soil particles.

Let us now discuss both in detail:

Bulk Density

The dry weight of soil per unit volume of soil is referred to as bulk density. It takes into account both the soil solids and the pore space. Bulk density of a soil is always lesser than that of particle density. As the soils tend to become finer from coarser, the bulk density decreases. Bulk density is usually expressed as g/cm^3 (Table 2.2).

Factors Affecting Bulk Density

1. Soil texture: As the soils tend to become finer from coarser, the bulk density decreases.
2. Organic matter: The higher the organic matter, lower is the bulk density, although the presence of organic matter in the soil provides stabilization to it.

Table 2.2 General guide for estimating moist bulk density (USDA)

Texture class	Bulk density (g cm^{-3})
Sand	1.65
Loamy sand	1.6
Sandy loam	1.55
Loam	1.5
Sandy clay loam	1.5
Silty clay loam	1.5
Silty loam	1.5
Clay loam	1.45
Silty clay	1.45
Sandy clay	1.4
Clay	1.35

3. Soil depth: The lower we go into the soil, the higher is the bulk density. Low organic matter and compaction brought on by the weight of the underlying layers are responsible for this. A soil's pore space is reduced, and its weight per unit volume is increased when it is packed.

Particle Density

The weight per unit volume of the solid portion of soil is termed as particle density or true density. Particle density varies from its counterpart bulk density with respect to pore space, as it does not take it into account.

Factors Affecting Particle Density

1. Heavy minerals: It is directly proportional to the presence of heavy minerals like magnetite, limonite, and hematite in the soil. The higher the heavy minerals, the higher will be the particle density.
2. Organic matter: With an increase in organic matter, the particle density decreases

2.4.5   Soil Color

The most evident and noticeable soil attribute is color (Foth, 1991). It is a crucial diagnostic feature for soil horizon demarcation and soil categorization because it reflects various soil characteristics and soil processes. Soil color is affected by various factors like mineral makeup, element concentration, amount of organic matter, and moisture content.

Significance of Soil Color (Mehra, 2004)

- Brown color is due to a mixture of organic matter and iron oxides.
- Clay is darker, and quartz is white or lighter in color. As a result, lighter colored soils contain more sand than darker colored soils. Soil with a darker color is therefore more fertile and clay-rich than soil with a lighter color.

- Within the water table fluctuation zone, poorly drained soils feature variegated colors of gray, brown, and yellow.
- Color of the soil may be inherited from its parent material, for example, red soils developed from red sandstone. This is called lithochromic color. The processes involved in soil formation can occasionally cause the soil to become colored. This is called genetic or acquired or pedochromic color, that is, red soils developed from granite, gneiss, or schist. In young soils, the parent material influences the color. The color of the soil tends to change from grayish to brownish or reddish when the soil develops to a later stage as a result of increased temperature and rainfall.
- The relationship between soil color and temperature is direct. Darker soils heated up faster because they absorb heat more easily. This isn't always the case, though, as certain soils hold more water and have higher levels of organic matter, which raises the soil's specific heat. As a result, although being darker in color, such soils could take longer to warm up.

Determination of Soil Color

Soil color is determined by Munsell Color Chart. The three variables that make up the Munsell color notations, which describe soil color, are as follows:

Hue: The major spectral color is hue, which is correlated with the dominant light wavelength. The hue notation of a color indicates its relation to red, yellow, green, blue, and purple. The letter abbreviation of each color in the rainbow (R for red, YR for yellow-red, and Y for yellow) is used as the symbol for hue, and it is followed by a number from 0 to 10. As the number rises, the tint becomes more yellow and less red.

Value: It is the lightness or darkness of the dominant spectral color. Numbers ranging from 0 for absolute black to 10 for absolute white are used to represent values.

Chroma: The dominant color's purity is measured by chroma: Numbers starting at 0 for grays and rising at regular intervals to a maximum of roughly 20 make up the chroma notation.

2.5 Soil Physics (B)

We will be talking about the physical characteristics of soil, such as soil air, soil temperature, and soil pH, in this portion of soil physics. All three are crucial to soil and have a significant impact on its biological and physical properties.

2.5.1   *Soil Air*

The gaseous state of the soil is called the soil atmosphere. The pores that are not filled with liquid are occupied by soil air. The soil air is the continuation of the atmospheric air, and from the soil pores into the atmosphere and from the atmosphere into the pore space, it is constantly moving.

Soil aeration is the process of gas exchange between the air in the atmosphere and the pore spaces of the soil. The respiration and survival of soil organisms and plant roots depend on soil aeration. This process prevents the toxicity of carbon dioxide evolved during respiration in the soil air as well as the deficiency of oxygen consumed during respiration of plant roots and soil microorganisms.

2.5.1.1   Composition of the Soil Air

Nitrogen, oxygen, carbon dioxide, water vapor, and other gases can be found in soil air. The respiration of roots and other organisms consumes oxygen while emitting carbon dioxide. As a result, soil air contains 10–100 times more carbon dioxide and slightly less oxygen than atmospheric air (Foth, 1991). Pressure differences between the two gases are created between the soil and the atmosphere. As a result, carbon dioxide diffuses out of the soil and oxygen diffuses in. Additionally, it has more water vapor than atmospheric air. Soil air has a nitrogen content that is almost identical to atmospheric air (Table 2.3).

2.5.1.2   Factors Affecting Composition of Soil Air

1. *Amount of air space*: The opportunity for gaseous exchange is greater in the top soil than in the subsoil because it has a greater amount of pore spaces than the subsoil. As a result, the top soil has more oxygen than the subsoil. The amount of pore space and consequently the soil aeration are influenced by the soil properties such as soil texture, bulk density, and aggregation.
2. *Organic matter*: Microbial activity is increased by the presence of organic matter that is easily decomposable. As a result, carbon dioxide production is increased. When carbon dioxide from the soil air is not quickly removed by gaseous exchange, its concentration rises.

Table 2.3 Composition of atmospheric and soil air

Component	Percent by volume	
	Atmospheric air	Soil air
Nitrogen	79	79–80
Oxygen	20.95	18–20
Carbon dioxide	0.03	0.15–0.3

3. *Soil compaction*: Poor aeration affects compact soils. Compared to surface soil, subsoil air has lower oxygen content and higher carbon dioxide content. Additionally, soil layers could get compacted to limit root growth.
4. *Soil moisture*: Immediately following a heavy downpour or irrigation, the macropores are filled with water and the oxygen content is zero. When the soil is artificially drained once more, the macropores are filled with air, raising the soil's oxygen content.
5. *Seasonal variation*: In temperate-humid regions, the soils are wet and cold in the spring, and there is poor gaseous exchange. The gaseous exchange will increase during the summer when the soils are dry. As a result, there will be little CO_2 and a relatively high amount of O_2.

2.5.1.3 Importance of Soil Aeration

1. *Plant and root growth*: Aeration of the soil is essential for normal plant growth. A sufficient supply of oxygen to roots and the removal of CO_2 from the soil atmosphere are critical for healthy plant growth. When the supply of oxygen is insufficient, plant growth either slows or stops completely, as accumulated CO_2 impedes the growth of plant roots.

   It is common in the root zone for O_2 to fall to 10% and CO_2 to rise to 10% without harming the plants. When soil pores fill with water, the vital oxygen is quickly depleted. Corn is extremely sensitive to this condition, but sorghum can withstand flooding for several days without suffering permanent damage (Eash et al., 2017).
2. *Activity of microorganisms*: Microorganisms in the soil require oxygen for respiration and metabolism as well. Some important microbial activities, such as organic matter decomposition, nitrification, sulfur oxidation, and so on, rely on oxygen in the soil air. The lack of oxygen (air) in soil slows the rate of microbial activity.
3. *Plant diseases*: Inadequate soil aeration also contributes to disease development. For example, citrus and peach dieback.
4. *Water and nutrient absorption*: It has been discovered that oxygen deficiency inhibits plants' ability to absorb nutrients and water. Plants that experience poor aeration (which can happen when the soil is waterlogged) show signs of water and nutrient shortage.

2.5.2 Soil Temperature

All plants require sunlight to grow. Sunlight provides the energy required for photosynthesis while also warming the soil and air in which crops grow. Soil temperature influences almost every physical, chemical, and biological activity that occurs in the soil. Soil temperature is frequently used to take management decisions, such as when to plant. Understanding how plants respond to climate changes is aided by understanding the flow of energy in the soil-plant-atmosphere system.

Temperature is an important soil property. The following points illustrate the importance of soil temperature:

- Below freezing, biological activity is extremely limited.
- Root elongation is inhibited by a soil horizon as cold as 5 °C.
- Temperature affects the chemical processes and activities of microorganisms.
- Soil expansion and contraction are caused by the alternate freezing and thawing of soils. This has an impact on rock weathering, structure formation, and plant root heaving (Foth, 1991).

2.5.2.1   Sources of Soil Heat

The primary source of energy for warming the soil is solar radiation (sunrays). The sun's rays are intercepted by dust particles, clouds, and other suspended particles. They either absorb, scatter, or reflect solar energy. Only a small portion of total radiation reaches Earth. Thermal energy from the sun is transmitted across space and through the atmosphere in the form of thermal infrared radiation. Other heat sources for soil include microbial decomposition of organic matter and respiration by soil organisms such as plants.

2.5.2.2   Factors Influencing Soil Temperature

- *Solar radiation*: Soil derives its heat from the sun. The amount of solar energy received by soil is determined by the composition of the atmosphere. Clouds, water vapor, and dust particles reduce the amount of solar energy that reaches the soil surface.
- *Position of the sun*: The sun is directly overhead in tropical areas. If the atmosphere is clear, 75% of solar radiation reaches the earth (Mehra, 2004). As one travels from the arctic to the equator, the temperature of the soil rises. Because the angle of incidence is greater in the polar regions, the amount of radiant energy received is much less than in the equatorial belt.
- *Slope of the land*: A surface that is perpendicular to the sun's rays receives more heat. If the same rays strike the earth's surface at right angles, they will cover a larger area. As a result, the same amount of heat is distributed over a larger area, and the energy received per unit area is proportionally reduced.
- *Vegetation*: Compared to vegetation-covered soils, bare soils warm up and cool down faster. Vegetative cover blocks the sun's rays, keeping the soil beneath cool. In the winter, vegetation acts as an insulator, reducing heat loss from the soil.

- *Soil factors*: This includes soil color, soil texture, soil depth, soil texture, soil structure, etc. For example, because of their higher water retention capacity, dark colored soils absorb more heat. Light colored soils reflect more heat back into the atmosphere. As a result, black and brown soils are warmer than gray soils.

2.5.2.3   Impact of Soil Temperature on Plant Growth

- *Seed germination and plant growth*: Changes in soil temperature have a significant impact on seed germination and root growth in most crops. Maize, for example, requires a soil temperature of 7–10 °C for germination and optimal root growth at around 25 °C. Root elongation is retarded at low temperatures. Plants' soil temperature requirements differ depending on the species. The temperature at which a plant thrives and grows the most is referred to as the optimum range (temperature). The term "growth range" refers to the entire temperature range in which a plant can grow, including the optimum range. The maximum and minimum temperatures at which a plant will die are referred to as survival limits.
- *Microbial activity*: Low temperatures slow metabolic activity and enzymatic reactions. Nitrification is hampered by high temperatures (above 35 °C). Most microorganisms thrive at temperatures between 25 and 35 °C.
- *Absorption of water and nutrient*: Low temperatures reduce nutrient availability, microbial activity, root growth and branching, and root growth and branching. Plants' ability to absorb nutrients and water decreases at low temperatures.
- *Plant diseases*: If germination is delayed due to low temperatures, there is a risk of fungi and bacteria damaging moist seeds.

2.5.3   Soil pH

- The concentration of hydrogen ions (H+) and hydroxyl ions (OH) in soil solution, expressed in moles per liter, is referred to as soil reaction. The pH of a system is a measurement of the concentration and activity of hydrogen ions (H+).
- It is defined as the negative log of the concentration of hydrogen ions, or $-\log ([H^+]) = \log 1/H^+$.
- The higher the concentration of H+, the lower the pH and the greater the acidity (Eash et al., 2017).
- The pH scale ranges from 1 to 14, with 7 being exactly neutral. This means that the concentrations of hydrogen and hydroxyl ions are equal at pH 7. As the concentration of hydrogen ions rises and that of hydroxyl ions falls, the pH falls below 7 and vice versa. Figure 2.5 depicts the pH ranges that might be encountered under natural soil conditions.

Soils are considered acidic if their pH falls below 5 and very acidic if their pH falls below 4. Soils, on the other hand, are considered alkaline above a pH of 7.5 and very

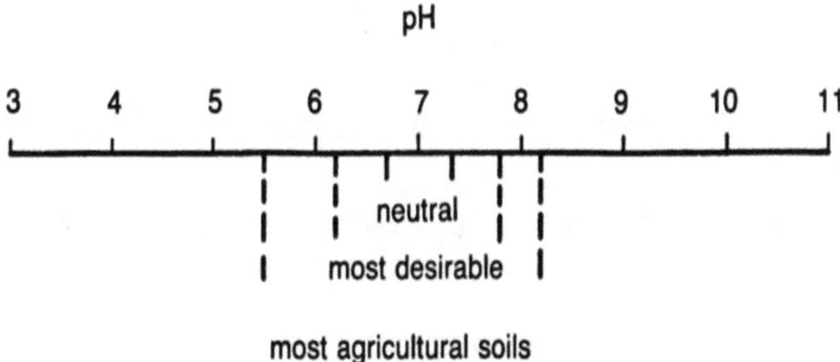

Fig. 2.5 Soil reaction is usually less than two pH units on either side of neutral (Eash et al., 2017)

alkaline above a pH of 8. Soils near the extreme ends of the pH scale are uncommon unless they have been contaminated by human activity.

2.5.3.1 Importance of Soil pH

Soil pH is an important chemical property that reflects a soil's overall chemical status. As a result, soil pH is the most commonly measured soil chemical property by both farmers and scientist. Its measurement can help determine how much lime or sulfur to add to the soil, which fertilizers to use (some fertilizers will acidify the soil while others will raise the pH), and whether a specific pesticide can be tank-mixed with a specific fertilizer. Soil pH is an important physicochemical property because it influences the following:

- Soil suitability for crop production
- Plant availability of soil nutrients
- Soil microbiological activity
- Soil liming and gypsum requirements
- Soil physical properties such as structure, permeability, and so on

2.5.3.2 Effect of Soil pH on Nutrient Availability

- The pH range of 6.5–7.5 is ideal for most crop plants. The general relationship between pH and nutrient availability is depicted in Fig. 2.6
- The maximum availability of N, P, K, Ca, Mg, and S occurs between pH 6.5 and 7.5.
- Minor element availability is higher in the acid range. Because the requirements for minor elements Zn, Fe, Mn, and Cu are so small, the quantities of these nutrients available at pH 6.5 to 7.5 are sufficient to meet the needs of plants.

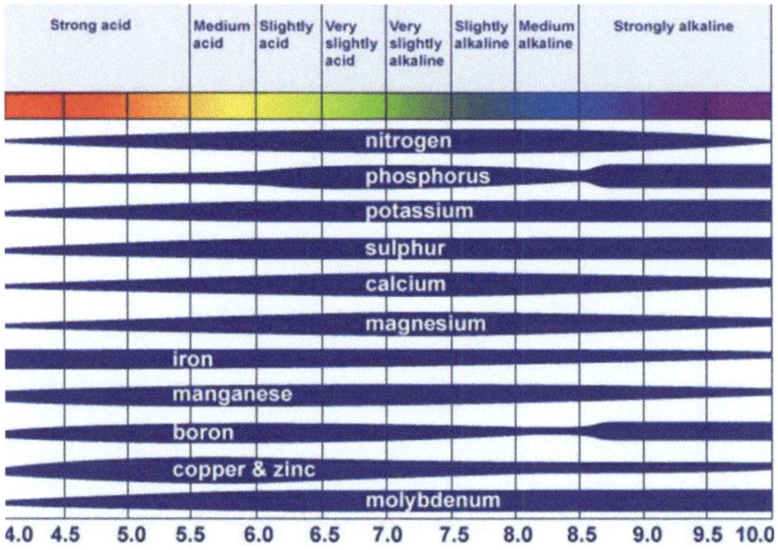

Fig. 2.6 Relationship between soil pH and availability of nutrients. (Research Review No. 78, S. Roques et al., 2013)

- As soil pH rises, molybdenum becomes more available. However, it is not deficient when the pH is low.

2.5.3.3 Effect of Soil pH on Soil Microorganisms

Bacteria and actinomycetes work better at intermediate and higher pH values (pH more than 6.0). However, the bacteria responsible for sulfur oxidation to sulfuric acid works properly at all pH levels. Because of the activities of these bacteria, sulfur can be used to create acidic conditions. Fungi also work satisfactorily at all pH levels.

Table 2.4 below shows the effect of pH on plant growth.

2.6 Soil Pollution and Its Mitigation

2.6.1 Introduction

The term "soil pollution" refers to the act of contaminating the soil with pollutants, hazardous substances, or other contaminants in such quantities as to degrade its quality and render it uninhabitable by organisms like insects and other microbes. It can also be defined as the addition of chemicals to the soil in toxic quantities to the environment and its inhabitants. This increase is primarily caused by human

Table 2.4 Effects of pH on plant growth (Colorado State University, CMG Garden Notes #222)

pH	Plant growth
>8.3	Too alkaline for most crops
7.5	Iron availability becomes a problem on alkaline soil
7.2	
7.0	6.8–7.2 near neutral
6.8	6.0–7.5 acceptable for most plants
6.0	
5.5	Reduced microbial activity
<4.6	Too acid for most plants

activities such as mining, modern agricultural practices, deforestation, indiscriminate dumping of human generated trash, and unregulated disposal of untreated wastes from various industries.

2.6.2 Point Source and Diffuse Soil Pollution

Soil pollution can be caused by both intentional and unintentional activities. These activities can include direct contaminant deposition into the soil as well as complex environmental processes that can result in indirect soil contamination via water or atmospheric deposition (Tarazona, 2014). Various types of soil pollution are discussed in the following sections.

2.6.2.1 Point Source Pollution

Soil pollution can be caused by a single event or a series of events within a specific area that release contaminants into the soil, and the source and identity of the pollution are easily identified. Point-source pollution is the name given to this type of pollution. The main sources of point-source pollution are anthropogenic activities. Former factory sites, insufficient waste and wastewater disposal, uncontrolled landfills, excessive agrochemical application, various types of spills, and other examples abound. Mining and smelting activities that use poor environmental standards are also sources of heavy metal contamination in many parts of the world (Lu et al., 2015; Mackay et al., 2013; Podolsk et al., 2015; Strzeboska et al., 2017).

2.6.2.2 Diffuse Soil Pollution

Diffuse pollution is pollution that spreads over large areas, accumulates in soil, and has no single or easily identified source. Diffuse pollution occurs when contaminants have been emitted, transformed, and diluted in other media before being transferred to soil (FAO and ITPS, 2015). There are numerous examples of diffuse pollution, which can include sources from nuclear power and weapons activities; uncontrolled waste disposal and contaminated effluents released in and near catchments; land application of sewage sludge; agricultural use of pesticides and fertilizers, which also add heavy metals, persistent organic pollutants, excess nutrients, and agrochemicals that are transported downstream by surface runoff; flood events; atmospheric transport; and deposition.

2.6.3 Causes of Soil Pollution

A soil pollutant is any factor that degrades the quality, texture, or mineral content of the soil or disrupts the biological balance of the soil's organisms. Soil pollution has a negative impact on plant growth.

Pollution in the soil is associated to the following:

- Modern agricultural practices
- Industrial and mining activities
- Lack of proper waste disposal
- Radioactive pollutants
- Acid rain

Modern Agricultural Practices
Synthetic chemicals are used to increase yield from limited land area in order to meet the increasing demand for food for an ever-increasing population. Pesticides and fertilizers have been widely used in recent decades, resulting in soil toxicity. They seep into the ground after mixing with water and gradually reduce soil fertility. Other chemicals alter the soil's composition and make it more prone to erosion by water and air.

Industrial and Mining Activities
Large numbers of industries that have sprouted up since the dawn of the industrial era without proper waste management systems have been the most significant contributors to soil pollution. Also, the amount of mining and manufacturing has increased, and the majority of industries rely on extracting minerals from the Earth. Whether it is iron ore or coal, the by-products are contaminated, and they are not disposed of in a safe manner. As a result, industrial waste that has been dumped on the soil surface for an extended period of time degrades it.

Lack of Proper Waste Disposal
Modern lifestyles, both urban and rural, generate massive amounts of waste, and a lack of waste management procedures exacerbates the problem of soil pollution. Urban wastes include both commercial and domestic wastes such as dried sludge and sewage, as well as garbage and rubbish materials such as plastics, glasses, metallic cans, fibers, and paper.

Radioactive Pollutants
Radioactive materials from nuclear testing lab explosions, radioactive fallout, and industrial sources of nuclear dust and radioactive waste seep into the soil and build up, causing soil pollution.

Acid Rain
When pollutants from the air mix with the rain and fall back to the ground, acid rain is the result. Some of the crucial nutrients in the soil may dissolve in the contaminated water, changing the soil's composition.

2.6.4   Pollutants in Soil

Pollutants are typically released into the environment as a result of human activity. Even though some substances and elements in soils occur naturally, human activities are the primary causes of soil pollution. Chemical characteristics have been used to categorize pollutants. The most common chemicals involved in causing soil pollution are petroleum hydrocarbons, heavy metals, pesticides, solvents, etc. Swartjes has proposed classification of pollutants, which is a useful tool for understanding them (Fig. 2.7).

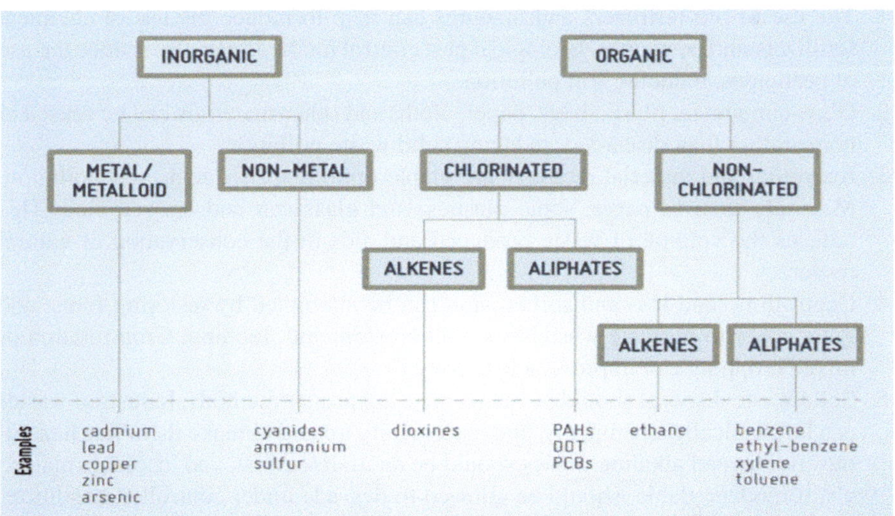

Fig. 2.7 Systematic categorization of main pollutants in soil. (Swartjes, 2011)

2.6.5 Effects of Soil Pollution

- *Agricultural*: It leads to reduced soil fertility, reduced nitrogen fertilizer, increased soil erosion, loss of nutrients, and reduced crop yield.
- *Industrial*: Ecological imbalance, release of pollutant gases, release of radioactive rays causing health problems, increased salinity, and reduced vegetation are some effects of soil pollution.
- *Humans*: Because we rely on the land for food, pollution from the soil is transferred to us in this way. Toxins accumulate in our bodies, causing chronic poisoning and leading to a variety of diseases. Reproductive health, birth and developmental defects, neurologic effects, malnutrition, and cancer-causing mutations in the body's cells are all on the rise today.
- *Plants*: Plants will be unable to adapt to sudden changes in the soil. Because of chemical changes, fungi and bacteria found in soils are unable to bind the soil, resulting in soil erosion. Large areas of land become barren, unable to support life. Toxins will be absorbed by the plants that do grow on these lands and transferred to the food chain.

2.6.6 Mitigation Measures

Soil pollution is a complex issue that necessitates collaborative efforts from governments, institutions, communities, and individuals. Some of the things we can do to prevent it are as follows:

1. The use of bio-fertilizers and manures can help to reduce the use of chemical fertilizers and pesticides. Biological pest control methods can also reduce the use of pesticides, reducing soil pollution.
2. Glass containers, plastic bags, paper, cloth, and other materials can be reused at home rather than discarded, reducing solid waste pollution.
3. Recycling and material recovery are viable options for reducing soil pollution. Materials such as paper, some plastics, and glass can and are recycled. This reduces the volume of waste produced and aids in the conservation of natural resources.
4. Controlling land loss and soil erosion can be attempted by restoring forest and grass cover to mitigate wastelands, soil erosion, and flooding. Crop rotation or mixed cropping can improve soil fertility.
5. Solid waste disposal should be managed using proper methods. Industrial wastes can be physically, chemically, and biologically treated to make them less hazardous. Acidic and alkaline wastes should be neutralized first, and insoluble materials, if biodegradable, should be allowed to degrade under controlled conditions before disposal.

2.7 Problematic Soils

Problematic soils are those that are unsuitable for field crop cultivation due to one or more unfavorable soil properties/characteristics (such as soluble salts, soil reaction, ESP, water logging, aeration, and so on) that have an adverse effect on optimum soil productivity. The problematic soils must be classified into different groups in order to develop special management systems for specific types of problems and constraints in crop production (Table 2.5).

2.7.1 Distribution

Salt-affected soils are found in arid and semiarid regions where evapotranspiration far outnumbers precipitation. Sodium, potassium, magnesium, calcium, chlorides, carbonates, and bicarbonates are examples of accumulated ions that cause salinity or alkalinity. Salt-affected soils are divided into three types: saline soil sodic and saline sodic soil. The following table shows the state-by-state distribution of salt-affected soils in India (Table 2.6).

Also, 49 million hectares (ha) of India's 157 million ha of arable land are acidic, with 26 million ha having soil pH values below 5.6 and the remaining 23 million ha having soil pH values between 5.6 and 6.5. Assam and portions of the states of UP, HP, WB, Bihar, Orissa, and TN have been found to have soils with low pH values (Table 2.7).

Table 2.5 Major problematic soils of India [Rajani (2019)]

Problematic soils	Key diagnosis	Major constraints
Acid soils	Soil pH less than 6.5	Fe, Al toxicity
Salt-affected soils		
1. Saline soils	Characterized by EC more than 4.0 dS	High osmotic potential, nutrient imbalance
2. Sodic soils	Characterized by EC less than 4.0 dS m^{-1}, pH more than 8.5, ESP more than 15.0.	Deteriorated physical condition, Na toxicity
3. Saline sodic soils	Characterized by EC more than 4.0 dS m^{-1}, pH more than 8.5, ESP more than 15.0	High osmotic potential, nutrient imbalance
Calcareous soils	$CaCO_3$ is greater than 5.0%	P, Fe deficiency

Table 2.6 Extent and distribution of salt-affected soils in India (Area × 1000 ha)

States	Waterlogged		Salt-affected area			
	Canal command	Total	Canal command	Outside canal	Coastal	Total
Andhra Pradesh	266	339	139	391	283	813
Bihar	363	363	224	176	Nil	400
Gujarat	173	484	540	327	302	1214
Haryana	230	275	455	Nil	Nil	455
Karnataka	36	36	51	267	86	404
Kerala	12	12	NA	NA	26	26
Madhya Pradesh	57	57	220	22	Nil	242
Maharashtra and Goa	6	111	446	NA	88	534
Orissa	196	196	NA	NA	400	400
Punjab	199	199	393	127	Nil	519
Rajasthan	180	348	138	984	Nil	1122
Tamil Nadu	18	129	257	NA	84	340
Uttar Pradesh	455	1980	606	689	Nil	1295
West Bengal	NA	NA	Nil	NA	800	800
Total	2190	4528	3469	3027	2069	8565

Tyagi, N.K.,CSSRI

Table 2.7 State-wise distribution of acid soils in India

Soil groups	pH range	Area (m ha)	States
Laterites soils	4.8–7.0	12.65	Mysore, MP, Eastern *Ghat* region of Orissa, WB, South Maharashtra, Kerala, Malabar Coast, Assam, parts of Santhal Parganas and Singhbhum in Bihar
Laterite and lateritic red soils	5.0–7.0	11.80	Kerala, Orissa, WB and Assam, parts of Santhal Parganas and Singhbhum in Bihar, Mysore, Bihar, MP, UP
Mixed red and black/yellow soils	5.5–6.5	23.66	Mysore, Bihar, MP, UP

2.7.2 Characteristics and Management of Problematic Soils

2.7.2.1 Saline Soils

Saline soils are those with a conductivity of saturation extract greater than 4 dS m^{-1} and an exchangeable sodium percentage less than 15. Typically, the pH is less than 8.5. Because of the surface crust of white salts, these soils were previously known as white alkali soils.

These soils are distinguished by saline efflorescence or white salt encrustation on the surface. These soils are known as "reh" in India and "th" in other countries.

Reclamation of Saline Soils

1. The basic principle in saline soil reclamation is the removal of excess salt to a desired level in the rooting zone.
2. The two most important components are leaching with high-quality water and adequate drainage.
3. The addition of organic matter improves the soil's physical conditions and increases its water holding capacity, which keeps the salt diluted.
4. During the first year of reclamation, rice is considered a satisfactory crop. Growing legumes is suitable for production after rice. Other management practices include the cultivation of salt-tolerant crops

   - High salt-tolerant crops: rice, sugarcane, and oats
   - Medium salt-tolerant crops: castor, cotton, and wheat
   - Low salt-tolerant crops: peas, pulses, and gram

2.7.2.2 Alkali or Sodic Soils

Alkali or sodic soil is defined as having a conductivity of saturation extract of less than 4 dS m^{-1} and an exchangeable sodium percentage greater than 15. The pH ranges from 8.5 to 10.0. Most alkali soils contain $CaCO_3$ in some form in the profile, particularly in arid and semiarid regions, and constant hydrolysis of $CaCO_3$ maintains the release of OH- ions in soil solution. The OH- ions released result in calcareous alkali soils maintaining a higher pH than noncalcareous alkali soils.

They are known as Usar in some parts of India and Kallar in others. These soils are mostly found in the arid and semiarid regions of Punjab, Uttar Pradesh, Bihar, and Rajasthan in India.

Management of Alkali Soils

1. Deep ploughing, land levelling, and bunding are required to improve the physical condition of the soil and make it pervious. Heavy irrigation is used after the addition of gypsum to aid in the leaching of Na-soluble salts.
2. Any reclamation measure must include the use of amendments and adequate leaching. There are several amendments available, including gypsum, S,H_2SO_4, $CaCl_2$, $FeSO_4$, and iron pyrites, but gypsum has been the most popular due to its low cost and ease of availability.
3. The addition of gypsum causes the formation of Na_2SO_4 in the soil, which is highly soluble and is easily leached away by heavy irrigation.
4. Irrigation water should be used correctly. It is known that as the amount of water in the soil decreases, the concentration of salts in the soil solution increases, so moisture should be kept at an optimum field capacity.
5. Organic matter addition is always beneficial in improving soil physical conditions such as aeration, water holding capacity, infiltration, and pulverization.
6. It is best to plant-tolerant and medium-tolerant crops.

- Tolerant crops include rice, sugar beet, and dhanicha.
- Wheat, barley, oats, and millets are medium-tolerant crops.
- Legume, maize, and groundnut are all sensitive crops.

7. Fertilizers with an acidic residual effect should be used, for example, urea.

2.7.2.3   Saline-Alkali Soil

A saline-alkali/sodic soil is one with a saturation extract conductivity greater than 4 dS m^{-1} and an exchangeable sodium percentage greater than 15. The pH varies and is usually higher than 8.5 due to the relative amounts of exchangeable sodium and soluble salts. The pH of soils dominated by exchangeable sodium will be greater than 8.5, while soils dominated by soluble salts will be less than 8.5.

Management of Saline Alkali Soils
Saline-sodic soil can be managed using the reclamation and management techniques advised for the reclamation of sodic soil.

2.7.2.4   Acid Soils

The presence of a higher concentration of H+ in soil solution and at exchange sites is referred to as soil acidity. They are distinguished by low soil pH and base saturation. The primary cause of acid soil formation is base leaching, though parent acidic rock also plays a role.

These soils have a high exchangeable Al3+ and H+ content and a pH less than 5.5 and respond well to lime application.

Acid soils have a negative impact on plant growth due to the presence of aluminum, manganese, and iron in toxic concentrations, a lack of calcium and magnesium, nutrient imbalance, and microbial imbalance.

About 47mha of land in the states of Assam, Tripura, Manipur, West Bengal, Bihar, Orissa, Karnataka, Tamil Nadu, Himachal Pradesh, and Kerala are covered by acid soils.

Management of Acid Soils
The management of acid soils should be aimed at increasing crop productivity, either through the addition of amendments to correct soil abnormalities or by manipulating agronomic practices based on climatic and edaphic conditions.

1. Soil amelioration: Lime has been recognized as an effective soil ameliorant because it reduces Al, Fe, and Mn toxicity while increasing acid soil base saturation, P, and Mo availability. Liming boosts atmospheric N fixation as well as N mineralization in acid soils by increasing microbial activity. The addition of lime raises the pH of the soil, thereby eliminating the majority of the major problems associated with acid soils. However, the economic feasibility of liming must be determined before making any recommendations.

2. Crop selection for acidity tolerance is an effective tool for combating this soil problem, and breeding such varieties is especially important for increasing productivity, especially in areas where liming is not economically feasible.

   - Arhar, soybean, cotton, and oats are among the sensitive crops.
   - Crops that are semi-tolerant include maize, sorghum, peas, wheat, and barley.
   - Tolerant crops include paddy, potato, tea, and millets.

2.7.2.5   Calcareous Soils

Calcareous soils contain enough free calcium carbonate ($CaCO_3$) to produce visible effervescence when treated with dilute 0.1 N hydrochloric acid. Calcareous soil has a pH greater than 7.0 and is classified as an alkaline (basic) soil.

Management of Calcareous Soils
- Because of the effect of soil pH on soil nutrient availability and chemical reactions that affect the loss or fixation of some nutrients, fertilizer management in calcareous soils differs from that in noncalcareous soils. $CaCO_3$ affects the chemistry and availability of nitrogen (N), phosphorus (P), magnesium (Mg), potassium (K), manganese (Mn), zinc (Zn), and iron (Fe). Copper (Cu) availability is also impacted.
- Acid-forming fertilizers, such as ammonium sulfate and urea fertilizers, sulfur compounds, organic manures, and green manures, are thought to be effective methods for lowering soil pH to a neutral pH value.

2.8   Soil Fertility

Soil fertility is defined as a soil's inherent ability to supply available nutrients present in it to plants in adequate amounts and proportions. Fertile soil is soil that has the ability to provide all of the plant nutrients to the plants in an adequate amount and in a balanced proportion for maximum growth and yield.

2.8.1   The Importance of Soil Fertility

The overarching principle of "feed the soil to feed the plant" serves as a guide for sustainable agricultural production. This fundamental idea is put into practice through a number of techniques meant to boost soil fertility and productivity by increasing organic matter, biological activity, and nutrient availability. The amount of organic matter in cultivated soils increases over time as a result of the addition of organic materials like composts, crop residues, and green manures. The soil's capacity to supply crops increases along with its level of organic matter. The ultimate objective is a nutrient-rich, biologically active, healthy soil with improved structure (Soil Fertility Management for Organic Crops, 2002, p. 1).

We cannot emphasize the importance of soil fertility enough. Approximately 700 million people worldwide, mostly in developing nations, go hungry every year (Cox, 1999, p. 11). Future prospects appear even less promising. About 65% of the region's rural population, including the majority of the poor, still depend primarily on farming on nonirrigated, rainfed lands (Prasad & Power, 1995, p. 8). The following section will highlight the importance of soil fertility:

- A crucial component of successful crop production is soil fertility, which is a measurement of the soil's ability to provide nutrients to plants. Fertilizers and soil fertility are two concepts that are very closely related. Fertilizer acts as a "SOURCE" from which we can continuously draw various nutrients and also add to the sink, whereas soil fertility acts as a "SINK" where plants can draw nutrients for maximum yield. In recent years, all nations have come to realize how crucial it is to manage fertilizer use and soil fertility in order to meet global demand for food and other agricultural raw materials.

- Intensive fertilizer use and cropping with high yielding varieties have undoubtedly increased food production and decreased food shortages, but they have also brought about a number of issues with soil fertility, soil and water pollution, and other environmental issues. On the other hand, rapid nutrient depletion brought on by overexploitation has been observed in many soils, along with widespread deficiencies of N, P, K, and S, as well as micronutrient deficiencies, particularly Zn and boron. Additionally, deforestation, shifting cultivation, burning of bushes, grasses, and cow dung, as well as soil erosion, degradation, and nutrient losses due to excessive fertilizer use, leaching losses, etc., have all contributed to the soil's declining fertility status. It is becoming clear that judicious and effective fertilizer use, as well as scientific soil fertility management, is key to the success of Indian agriculture in the future.

- Soil fertility issues cannot be solved simply by providing plant food nutrients; effective management is also essential, as fertilizer is one of the most expensive inputs. To maximize returns with the least amount of investment, fertilizer scheduling must be well balanced. Aside from fertilizers, farmers are unable to apply enough organic manure due to a lack of biomass resources. Sound soil and crop management practices, judicious use of fertilizers, and integrated nutrient management practices must therefore be adopted to improve and maintain good soil fertility and better soil physical condition for the purpose of sustained crop production.

2.8.2 Factors Affecting Soil Fertility

There are two main factors that affect the soil fertility. They are categorized as natural factors and artificial factors.

Natural Factors The soil structure is impacted by these elements. Their impact begins even before the soil is formed, which are as follows:

- Parent material: Rock weathering results in the production of soil. The soil that is created from rock will be fertile if the source material contains significant amounts of plant nutrients. The soil fertility of kaolinite clay mineral-based soils is quite poor. Sedimentary rock-derived soils are more productive
- Depth of soil profile: In comparison to other soils, such as sandy soils, soils with a deep soil profile are more fertile.. It is because plants have deep, extensive roots that may take up a lot of water and nutrients from the earth.
- Age of the soil: Old soils lose nutrients as a result of weathering, erosion, leaching, and crop growth. Soils that have just been developed are therefore more fertile than those that have been around for a while.
- Topography: As a result of erosion and leaching, nutrients that accumulate at lower levels are lost from the soils of upper regions. Therefore, compared to the soils of plains and lower regions, the fertility of muddy and mountainous terrain is lower.
- Climate: Moisture, rain, temperature, and air are the four basic elements that make up a climate. Due to leaching at high temperatures and rapid organic matter decay, plant nutrients move from the upper surface of the soil to the bottom layer. Thus, soil fertility is reduced. The higher soil particles are swept away by the wind, reducing the soil's fertility.

Artificial Factors
- Water logging: Long-term flooding of a soil with excess water is detrimental to the soil's physical characteristics. Owing to poor aeration caused by water logging, aerobic microorganism activity is decreased, pathogenic microorganisms become active, soluble plant nutrients decrease due to leaching, and soil fertility is negatively impacted.
- Cropping system: The same nutrients are used more, and the soil becomes deficient in them, making it less fruitful, if the same crop is cultivated on a field continuously for a number of years. Therefore, it is not advisable to grow the same crops in the same soil for an extended period of time. Science-based soil management is required for the crops.
- Weed control: Weeds (unwanted plants) are comparable to plants in that they make use of nutrients, moisture, light, and air. The soil's fertility for the primary crop is reduced if these weeds are left unchecked since they use up a lot of nutrients in their growth.
- The use of manures and fertilizers: Manures and fertilizers boost the fertility of the soil by supplying plant nutrients to it. In order to improve the soil's fertility, specialized fertilizers and manures need to be introduced.

2.8.3   Soil Fertility Evaluation

Soil fertility evaluation is the diagnosis of the soil's nutrient status using various techniques or methods. There are three fundamental tools for assessing soil fertility. They are detailed below.

(a) Visual symptoms of nutrient deficiency
(b) Plant tissue analysis
(c) Soil analysis
(a) *Visual Symptoms of Nutrient Deficiency*

It is a qualitative measure of plant nutrient availability. It is a visual method of assessing soil fertility and diagnosing plant diseases. A lack of one or more nutrient elements may result in an abnormal appearance of the growing plant. The appearance of deficiency symptoms on plants has long been used as an indicator of soil fertility.

If a plant is deficient in a particular element, various symptoms may appear. This visual method of soil fertility evaluation is very simple and inexpensive and does not require complicated equipment; however, judging the deficiency symptoms becomes difficult when many nutrients are involved. In such cases, only an experienced individual can make sound decisions.

(b) *Plant Tissue Analysis*

Plant tissue analysis is the determination of the total elemental content of plants or specific plant parts in a laboratory (Steinhilber & Salak, 2010; Reuter & Robinson, 1997). It is used for a variety of purposes, including crop nutrient status monitoring and troubleshooting. It is also used to make nutrient recommendations for perennial fruit crops (Steinhilber & Salak, 2010). It is the only way to determine whether or not a crop has received adequate nutrition during the growing season (Flynn et al., 2004).

(c) *Soil Testing*

A soil test is a chemical method for estimating a soil's nutrient-supplying power. It is much faster and has an advantage over other methods of determining soil fertility. Before planting a crop, one can assess the soil's requirements. A soil test measures a portion of the soil's total nutrient supply. Soil testing is important in today's modern and intensive agricultural production system because it involves the continuous use and misuse of soil without proper care and management. Soil analysis is useful for gaining a better understanding of the soils in order to increase crop production and achieve a sustainable yield. Soil testing is an essential tool in soil fertility management (Fig. 2.8).

Fig. 2.8 Visual symptoms of selected nutrient deficiencies in Maize. (Food and Fertilizer Centre 2001; Nutrico 2011; 12th Publication Agriculture food and Rural Affairs 2007)

MCQs

1. Which soil is most widely available soil in India?

   - Black Soil
   - Red Soil
   - Alluvial Soil
   - None of the above

2. Red soil covers _______ of the Indian landmass

   - 20%
   - 16.5%
   - 9%
   - 18.5%

3. Red soil is red in color because of presence of?

   - Copper Sulfate
   - Ammonium Nitrate
   - Ferrous Oxide
   - Calcium Carbonate

4. _____ is highly acidic with low humus content

   - Red soil
   - Laterite soil
   - Black soil
   - Alluvial soil

5. The degradation of clay minerals in the strong acid soil is called ________

   - Illuviation
   - Salinization
   - Podzolization
   - Calcification

6. How many types of soil structures are found on earth?

   - 3
   - 7
   - 6
   - 4

7. Which of the following giving crop is medium salt-tolerant?

   - Cotton
   - Rice
   - Gram
   - Peas

8. Alkali soils are also known by the name _____ in some parts of India

   - Ker
   - Usar
   - Reh
   - All of the above

9. Saline soils are characterized by EC more than?

   - 4.0 dS
   - 0.5 dS
   - 8.0 dS
   - 6.0 dS

10. *Reh* is another name for ___ soils

    - Alkali
    - Saline
    - Acid
    - None of the above

References

Clothier, B. E., Green, S. R., & Deurer, M. (2008). Preferential Flow and Transport in Soil: Progress and Prognosis. *European Journal of Soil Science, 59*(1), 2–13.

Cox, G. W. (1999). *Alien species in North America and Hawaii: Impacts on natural ecosystems.* Island Press.

Eash, N. S., Sauer, T. J., O'Dell, D., & Odoi, E. (2017). *Soil science simplified* (6th ed., p. 7). John Wiley & Sons.

FAO and ITPS. (2015). Status of the World's Soil Resources (SWSR)—main report. In *Food and agriculture organization of the United Nations and intergovernmental technical panel on soils.* FAO.

Flynn, N. J., Bradley, C., & Tebbett, I. R. (2004). The role of environmental Management Systems in Agriculture. *Environmental Science & Policy, 7*(1), 47–54.

Food and Fertilizer Centre 2001; Nutrico 2011; 12th Publication Agriculture food and Rural Affairs 2007.

Foth, H. D. (1991). *Fundamentals of Soil Science, 3*, 22–23.

https://gcwgandhinagar.com/econtent/document/15876181510EVSAECC01_soil%20pollution.pdf

https://www.researchgate.net/post/what_kind_of_agicultural_chemicals_are_creating_soil_pollution

Lu, C., Tian, H., Liu, M., Ren, W., Xu, X., Chen, G., & Zhang, C. (2015). Modeling the effects of land use and land cover change on nitrogen cycling and carbon dynamics in terrestrial ecosystems: A case study in China. *Ecological Modelling, 312*, 105–116.

Mackay, D., Paterson, S., Di Guardo, A., & Cowan, C. E. (2013). Evaluating the environmental fate of chemicals using a level III fugacity model: Application to organic chemicals. *Chemosphere, 22*(4), 511–522.

Mehra, R. K. (2004). *Textbook of Soil Science, 9*, 114–115; 124; 12: 163, 173.

Pidwirny, M. (2006). *Fundamentals of physical geography* (2nd ed.). Retrieved from http://www.physicalgeography.net/

Podolsk, V., De Marchi, M., Dal Zotto, R., & Reggiani, M. (2015). Sustainable soil management in the agro-ecological landscape of Italy: The role of erosion control measures. *Journal of Soil and Water Conservation, 70*(6), 476–485.

Prasad, R., & Power, J. F. (1995). *Nutrient Management for Sustainable Agriculture.* CRC Press.

Rajani, A. (2019). *Problematic soils and their management 3; 16.* https://doi.org/10.13140/RG.2.2.13993.75360

Reuter, D. J., & Robinson, J. B. (1997). *Plant analysis: An interpretation manual.* CSIRO Publishing.

Steinhilber, M., & Salak, B. (2010). *Soil fertility management: A guide to evaluating soil fertility and nutrient status.* University of Maryland Extension.

Strzeboska, M., Zgłobicki, W., & Kozyra, J. (2017). Assessing soil erosion risk in Poland: The use of RUSLE model and geo-environmental information systems. *Geomorphology, 280*, 17–28.

Swartjes, F. A. (2011). *Dealing with contaminated sites: From theory towards practical application.* Springer Science & Business Media.

Tarazona, J. V. (2014). Ecotoxicology in Latin America: An overview of challenges and advances. *Environmental Science and Pollution Research, 21*(10), 6052–6061.

USDA General Guide for Estimating Moist Bulk Density. https://www.nrcs.usda.gov/wps/portal/nrcs/detail/soils/survey/office/ssr10/tr/?cid=nrcs144p2_074844

Chapter 3
Crop Physiology

Abstract This chapter of the book focuses on various aspects of crop physiology, that is, the study of functions and vital processes occurring in a plant. It is an integrative science applying knowledge from various disciplines for crop management problems and has an important role to play in increasing yield that will be needed to feed the burgeoning world population. It caters to all the basic processes, which occur in a within the plants. Topics covered in the chapter include water absorption, mineral nutrition, respiration, photosynthesis, photorespiration, evapotranspiration, and many others.

Keywords Water potential · Glycolysis · Photorespiration · Imbibition · Guttation · Transpiration pull

3.1 Introduction to Crop Physiology

We all know that plants are living things that carry out all of the fundamental biological functions like utilizing and absorbing water, producing and assimilating energy, etc. These are all referred to as vital plant processes. These internal processes and their functional facets are studied under crop physiology.

According to Noggle and Fritz (1983), it is the science concerned with processes and functions, the responses of plants to environment, and the growth and development that results from the responses. It helps to understand various biological processes of the plants like photosynthesis, respiration, transpiration, translocation, nutrient uptake, plant growth regulation through hormones, and such other processes, which have profound impact on crop yield.

L. Ahmad et al., *Fundamentals and Applications of Crop and Climate Science*,
https://doi.org/10.1007/978-3-031-61459-0_3

3.2 Importance of Crop Physiology

In-depth study of crop physiology can help with many elements of agriculture and offer solutions that are applicable to not only agriculture but all its allied sciences. Understanding the physiological processes involved in seed germination, seedling development, crop establishment, vegetative growth, flowering, plant hormone interaction, crop maturity, fruit and seed setting, nutrient physiology, stress (biotic/abiotic) physiology, etc. provides a sound scientific foundation for efficient monitoring and favorable manipulation of these phenomena.

Thus, a basic scientific foundation for comprehending many aspects of metabolism, growth, and development is provided by an understanding of the physiology of crop plants. This is crucial for crop improvement, in terms of yield and monetary benefits, and technological advancement in agriculture.

3.3 Overview of a Plant Cell

The smallest unit of life in all species is the cell. Plants are made up of a variety of cells much like other organisms are. The plant cell is enclosed by a cell wall, which is responsible for giving the plant cell shape. Other organelles exist in addition to the cell wall and are connected to various cellular functions, all important for proper growth and development of the plant.

The primary building block of life in species belonging to the kingdom Plantae is plant cells. These cells are eukaryotic in nature and hence have a true nucleus as well as specialized components called organelles that serve a variety of purposes. While bacteria and archaea have simpler prokaryotic cells, animals, plants, fungus, and protists have more complex eukaryotic cells. The cell walls, plastids (chromoplasts and chloroplasts), plasmodesmata, and central vacuole of plant cells distinguish them from the cells of other species.

Plant cells are typically between 10 and 100 micrometers length. Plant cells produce their own food through photosynthesis, in contrast to animal cells. In this process, oxygen and glucose are produced from water and carbon dioxide taken from surroundings, in the presence of light as the energy source (Fig. 3.1).

3.3.1 Water Absorption

Plant roots absorb water from the soil and use it for biochemical processes such as photosynthesis. Water absorption is of two types, passive and active (Renner, 1912).

1. *Active absorption*: The root cells actively participate in water absorption during this process, and metabolic energy released during respiration is consumed. It is of two kinds:

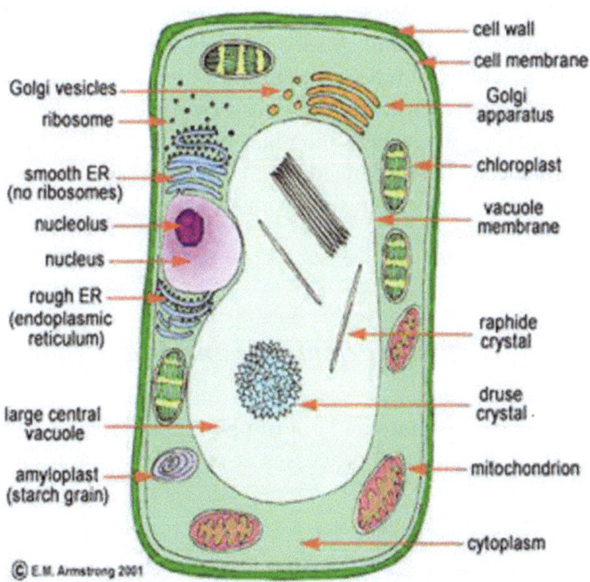

Fig. 3.1   Structure of a typical plant cell. (Armstrong, 2001)

- Osmotic active absorption: Osmotic absorption is the process by which water is absorbed from the soil into the xylem of the roots along an osmotic gradient.
- Non-osmotic active absorption: Non-osmotic absorption occurs when water is absorbed against the osmotic gradient

2. *Passive absorption*: The use of metabolic energy is not required for this type of water absorption. Passive water absorption occurs when the rate of transpiration is high. The rapid evaporation of water from the leaves during transpiration causes tension in the leaves' xylem. This tension is transferred to water in the xylem of the roots via the xylem of the stem, and the water rises to reach the transpiring surfaces. As a result, soil water enters the cortical cells via root hairs to reach the xylem of roots and maintain the water supply. Because the force for this water entry is created in the leaves by rapid transpiration, the root cells remain passive during this process.

3.3.2   *Plant-Water Relations*

All living cells contain a significant amount of water, which serves as a medium in which all substances are dissolved and go through a variety of reactions. Water itself is both a reactant and a product. It serves as a mode of transportation. Water is necessary for cellular growth and turgidity. Watermelon has 92% water, while the majority of herbaceous plants have 85–90%. In trees and shrubs, the woody parts contain a comparatively smaller amount of water compared to the softer parts, which contain a significant amount. A seed may appear dry, but it contains 10–15% water. It will stop breathing and die without this water.

A plant depends heavily on water for its survival. A total of 500 g of water are absorbed by the roots, transported to the plant body, and lost to the atmosphere for every gram of organic matter produced by the plant. Water deficiency and severe cellular process malfunction can result from even minor imbalances in this flow of water.

3.3.3   A Brief of Diffusion, Osmosis, and Imbibition

The movement of substances into and out of plant cells occurs as a gas or solution. There are typically three physical processes at play. They are imbibition, osmosis, and diffusion.

3.3.3.1   Diffusion

Diffusion is the movement of particles or molecules from an area of higher concentration to an area of lower concentration. Gases diffuse more quickly than liquids or solutes do. The pressure exerted by the diffusion particles, which is known as the diffusion pressure, is directly proportional to their concentration. Diffusion always occurs along a diffusion pressure gradient, moving from a region of higher diffusion pressure to a region of lower diffusion pressure.

The process of diffusion is crucial to the survival of plants.

1. It is a crucial step in the exchange of gases during photosynthesis and respiration.
2. Diffusion is important in stomatal transpiration, where water vapor diffusion from the intercellular space into the outer atmosphere occurs through open stomata.
3. During passive salt uptake, ion absorption takes place by diffusion.

3.3.3.2   Osmosis

Osmosis is the process of solvent molecules diffusing into a solution through a semipermeable membrane (sometimes called as osmotic diffusion). The diffusion of solvent from the less concentrated solution into the more concentrated solution will occur if two solutions with different concentrations are separated by a semipermeable membrane until both solutions reach an equal concentration.

Osmotic Pressure   When an osmotically active solution is separated from its pure solvent by a semipermeable membrane under ideal osmosis conditions that do not permit solution dilution, the maximum pressure that can develop in the solution is known as the osmotic pressure. It is proportional to the concentration of dissolved solutes in the solution and is measured in atmospheres, bars, or Pascal's. Higher OP solution has higher concentration. A solution's OP is always greater than that of its pure solvent. The movement of solvent molecules occurs during osmosis from the solution with the lower osmotic pressure (i.e., less concentration as hypotonic) into the solution with the higher osmotic pressure (i.e., more concentrated as hypertonic). If the two solutions separated by the semipermeable membrane are of equal concentration and have equal osmotic pressures, osmotic diffusion of the solvent molecules will not occur (i.e., they are isotonic).

Endosmosis and Exosmosis
1. Endosmosis is the process by which water enters the cell sap of a living plant cell by osmosis when the plant cell is immersed in water or a hypotonic solution with an OP lower than the cell sap. Water enters the cell sap; as a result, creating pressure that presses the protoplasm up against the cell wall and causes it to become turgid. Turgor pressure is the name given to this pressure.
2. The plant cell becomes flaccid if, on the other hand, it is placed in a hypertonic solution (whose OP is higher than the cell sap). This occurs because water leaks from the cell sap into the outer solution. Exosmosis is the name for this process. In an isotonic solution, cells or tissues will continue to exist as such (Fig. 3.2).

3.3.3.3 Imbibition

Certain substances absorb and swell when placed in a specific liquid. For example when some grass, dry wood, or dry seeds are placed in water, they absorb the water quickly and swell significantly, increasing their volume. These substances are referred to as imbibants, and the phenomenon is referred to as imbibition. Imbibate is the liquid that is imbibed.

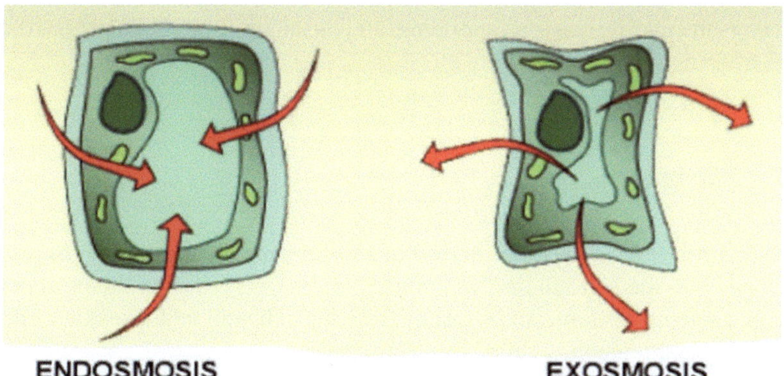

Fig. 3.2 Diagrammatic representation of endosmosis and exosmosis

Imbibition is critical to the survival of plants. The imbibition of water by the cell walls of the root hairs is the first step in the absorption of water by the roots of higher plants. Dry seeds require water by imbibition to germinate.

As a result of imbibition, a pressure known as imbibition pressure or matric potential is created.

3.3.3.4   Diffusion Pressure Deficit (DPD)

The reduction in the diffusion pressure of water in a solution over its pure state is called diffusion pressure deficit or DPD (Meyer, 1938). It was used in place of water potential and is an older term. The diffusion pressure of pure water is maximum and gets lowered by addition of solute particles in it. Due to the presence of DPD, a solution will always tend to make up the deficit by absorbing water. It is also called as suction pressure.

3.3.3.5   Water Potential (Slatyer & Taylor, 1960)

Every component of a system contains free energy that can be used to perform work under constant temperature conditions. Chemical potential is the term used for free energy/mole in nonelectrolytes. When it comes to water, the chemical potential of water is referred to as water Potential. A Greek letter Psi represents the chemical potential. The water potential for pure water is zero. The presence of solute particles reduces the free energy or water potential. As a result, the water potential of a solution is always less than zero, indicating a negative value.

3.3.3.6   Osmotic or Solute Potential

It is the decrease in the chemical potential of pure water caused by the presence of solute particles in it. Solute particles reduce the free energy of water by diluting it, increasing entropy, decreasing vapor pressure, raising the boiling point, and decreasing the freezing point.

3.3.4   *Plasmolysis*

When a plant cell or tissue is immersed in a hypertonic solution, water expels from the cell sap into the surrounding solution via exosmosis, causing the protoplasm to shrink or contract. Protoplasm separates from the cell wall and takes on a spherical shape. This phenomenon is called as plasmolysis.

Incipient plasmolysis is the stage at which protoplasm begins to contract from the cell wall. Endosmosis occurs when a plasmolyzed cell in tissue is placed in water. Water enters the cell sap, the cell becomes turgid, and the protoplasm returns to its normal shape and position. This is known as deplasmolysis.

3.4 Mineral Nutrition

An inorganic ion that must be obtained from the soil and is necessary for plant growth is generally referred to as a mineral nutrient. Nutrients are the chemical form of elements that are applied to plants. The provision and uptake of chemical compounds necessary for plant growth and metabolism can be referred to as nutrition. Three sources, namely, the atmosphere, water, and soil, provide the nutrients necessary for the growth and development of higher plants.

3.4.1 *Methods to Study the Mineral Nutrition in Plants*

Julius von Sachs, a prominent German botanist, demonstrated for the first time in 1860 that plants could be grown to maturity in a defined nutrient solution without soil. This formed the basis of modern-day hydroponics, which is a method of growing plants using mineral nutrient solutions (Sardare, 2013). Since then, a variety of improvised methods have been used to try to determine the mineral nutrients required by plants. The essence of all of these methods is plant culture in a soil-free, defined mineral solution. These methods necessitate the use of purified water and mineral nutrient salts.

A mineral solution suitable for plant growth was obtained after a series of experiments in which the roots of the plants were submerged in nutrient solutions and where an element was added, substituted, removed, or given in varying concentrations. This process allowed for the identification of essential elements and the detection of their deficiency symptoms. For the successful commercial production of vegetables like tomato, seedless cucumber, and lettuce, hydroponics has been used as a technique.

There were nine mineral nutrients total in *Sachs'* nutrient solution (K, N, P, Ca, S, Na, Cl, Fe, and Mg). Because they were supplied in the form of carbon dioxide and water and were not regarded as mineral elements, carbon, hydrogen, and oxygen were not included in this total.

3.4.2   Essential Elements

Arnon and *Stout* coined the term essential mineral element *(1939)*. These are the macro- and microelement compositions; in the absence of any of these elements, the plant cannot maintain normal growth and develops deficiency symptoms, affects metabolism, and dies prematurely.

Only 17 of the many elements discovered in plant tissues are required by all higher plants (nickel being the recently enlisted). C, H, O, N, P, K, Ca, Mg, S, Zn, Cu, Fe, Mn, B, Cl, and Mo are the elements. In the absence of any of the essential elements, plants develop deficiency symptoms and die prematurely.

3.4.2.1   Criteria for Essentiality (*Arnon and Stout*)

They determined that an element must meet the following three criteria in order to be considered essential.

1. In the absence of the mineral element, a plant must be unable to complete its life cycle.
2. The element's function must not be replaced by another mineral element.
3. The element must be directly involved in plant metabolism, such as a component of an essential plant constituent like an enzyme, or it must be required for a specific metabolic step like an enzymatic reaction.

The second criterion has some exceptions; it cannot be followed as written because some elements can be replaced by others without harming the plant. The monovalent cation K+, for example, can be replaced by Na+; both of these ions are important in osmoregulation.

3.4.2.2   Classification of Essential Elements

- Macronutrients

Major nutrients or macronutrients are nutrient elements that are required for plant growth in relatively large quantities. C, H, O, N, P, K, Ca, Mg, and S are the major elements required for plant growth. C, H, and O are among the nutrients that plants absorb from the atmosphere and water. The elements N, P, K, Ca, Mg, and S are taken up by plants from the soil and applied as chemical fertilizers through the soil or foliage.

- Micronutrients

Minor or micronutrients or trace elements are nutrient elements that are required in relatively small quantities. Zn, Cu, Fe, Mn, Mo, B, and Cl are the micronutrients required for plant growth.

In addition to the 17 essential elements, there are some beneficial elements such as sodium, silicon, cobalt, and selenium. They are required by higher plants.

3.4.2.3   Role of Macro- and Micronutrients

Both macro- and micronutrient elements can be classified into four groups for ease of understanding (Malik & Srivastava, 1982):

Group 1: N and S are covalently bonded organic matter constituents that are present in reduced form.

Group 2: P, B, and Si are found as oxyanions such as phosphate, borate, and silicate.

Group 3: K, Na, Mg, Ca, and Cl are involved in osmoregulation and ionic balance, as well as specific enzyme conformation and catalysis functions (e.g., metallo-protein complexes).

Group 4: Fe, Cu, Mo, and Zn are present as structural chelates or metalloproteins and are also involved in oxidation-reduction (redox) reactions (first three elements).

Table 3.1, mentioned below, provides a summary of physiological functions and mineral element deficiency symptoms in plants.

3.4.3   Mechanism of Mineral Uptake

Mineral nutrients can be found as soluble fractions of soil solution or as adsorbed ions on colloidal particle surfaces. There are two broad categories of theories proposed to explain the mechanism of mineral salt absorption:

 I. Passive absorption
II. Active absorption

- *Passive absorption*: When the concentration of mineral salts in the outer solution is greater than the concentration in the cell sap of the root cells, the mineral salts are absorbed according to the concentration gradient via a simple diffusion process. Because it does not necessitate the expenditure of metabolic energy, this is referred to as passive absorption.

- Mechanism of passive absorption

  (i) Diffusion: It occurs when small, nonpolar molecules diffuse across membranes (i.e., O_2, CO_2). Ions or molecules move from a higher concentration to a lower concentration during this process. It does not require any energy.

  (ii) Facilitated diffusion: Specific proteins in the membrane facilitate the diffusion of small polar species (such as H_2O, ions, and amino acids) down the electrochemical gradient. This process is known as facilitated diffusion.

Table 3.1 Physiological functions and deficiency symptoms of mineral elements in plants

Elements	Primary function	Specific deficiency symptoms
Oxygen	Final electron acceptor in aerobic respiration; element in carbohydrates, nucleic acids, and many other organic compounds	
Carbon	Building block of all organic compounds in the plants' body	
Hydrogen	Component of water and all organic compounds	
Nitrogen	Constituent of amino acids, proteins, chlorophyll, nucleic acid, some coenzymes	"V"-shaped chlorosis starting from the tip in lower leaves, plant becomes pale green, poor growth
Potassium	Involved in enzyme activation, protein metabolism, cell membranes, ionic balance, cell extension growth, opening and closing of stomata, cell turgor	Chlorosis of older leaves later turns into necrotic lesions on leaf tip, rolling of leaves, shortening of internodes leading to stunted growth
Calcium	Constituent of cell wall, middle lamella, enzyme cofactor, controls cell permeability and cell wall configuration, involved in cell signaling	Twisting and deformation of growing tips and youngest leaves, later necrosis occurs at the leaf
Phosphorus	Involved in photosynthesis, constituent of high-energy intermediaries like ATP, nucleic acids, phospholipids in membranes	Stunted growth, foliage turns dark green, high root/shoot ratio, delayed maturity of plants
Magnesium	Central element of chlorophyll molecule and an activator of several enzymes	Interveinal chlorosis of older leaves, purple coloration with necrotic spots, leaves become stiff and intercostal veins twist
Sulfur	Component of some amino acids (cysteine and methionine) and coenzymes	Symptoms appear in youngest leaves, general chlorosis of the entire leaf including vascular bundles
Iron	Involved in chlorophyll biosynthesis, structural component of cytochromes and ferredoxin	Symptoms appear first in younger growing organs, interveinal chlorosis; in severe cases, leaves become white which later dries out
Chlorine	Stomatal regulation in some plants (e.g., onion), stimulation of proton pumping ATPase located at tonoplast, involved in photosynthetic oxygen evolution, osmoregulation	Symptoms appear first in younger leaves, reduction in leaf surface area, wilting of leaf margins, interveinal chlorosis of mature leaves, highly branched root system
Copper	Activator of some enzymes like Cu-Zn SOD	Young leaves turn dark-green, twisted with necrotic spots, depressed internode growth, bushy appearance, stunted growth, reduction in panicle formation
Manganese	Activator of some enzymes (Mn-SOD) involved in PS I as water-splitting complex	Small yellow spots and interveinal chlorosis on younger leaves

(continued)

Table 3.1 (continued)

Elements	Primary function	Specific deficiency symptoms
Zinc	Activator of some enzymes, involved in the synthesis of auxin	Stunted growth due to shortening of internode called "rosette," drastic decrease in leaf size
Molybdenum	Cofactor of enzymes involved in nitrogen metabolism	Chlorosis and stunted growth in younger leaves, drastic reduction in size, whiptail appearance of leaf blade
Nickel	Metal cofactor of urease enzyme	Symptoms closely related to Fe deficiency

Modified after Malik and Srivastava (1982)

> (a) Channel proteins: Specific proteins in the membrane form channels (channel proteins), which can open and close and allow ions or H_2O molecules to pass in single file at extremely fast rates. The same facilitated diffusion process is used by a K+ and NH_4+ channel. This process can also allow Na+ into the cell.
>
> (b) Transporters or co-transporters or carriers: Transporters or co-transporters are responsible for the transport of ions and molecules across membranes in another mechanism. In contrast to channel proteins, transporter proteins bind only one or a few substrate molecules at a time. After binding a molecule or an ion, the transporter undergoes a structural change that is unique to that ion or molecule. As a result, the rate of transport across a membrane is slower than that of channel proteins.

- *Active absorption:* It refers to the absorption of minerals with the direct expenditure of metabolic energy

- <u>Mechanism of active absorption</u>

- Larger or more charged molecules have a difficult time crossing a membrane, necessitating active transport mechanisms (i.e., sugars, amino acids, DNA, ATP, ions, phosphate, proteins, etc.). Active transport occurs across a selectively permeable membrane via ATP-powered pumps that move ions against concentration gradients. This mechanism makes use of the energy released by ATP hydrolysis.

- The Na+-K + ATP pump transports K+ into the cell and Na+ out. The Ca+2 -ATP pump is another example.

3.4.4 Translocation of Mineral Nutrients

- The bulk transfer of water and dissolved minerals from one part of the plant to another via the xylem and phloem is referred to as translocation.

- Mineral salts are transported through the xylem with the ascending water flow. A small amount of material exchange occurs between the xylem and phloem.
- Mineral ions such as phosphorus, sulfur, nitrogen, and potassium are stored in senescing parts such as old leaves. These are frequently remobilized and transferred to younger leaves. Some structural elements, such as calcium, are not remobilized.
- The majority of nitrogen, phosphorus, and sulfur are transported in organic form as amino acids and related compounds. Some nitrogen is translocated as inorganic ions as well.

3.5 Photosynthesis

Photosynthesis is a vital physiological process in which the chloroplast of green plants synthesizes sugars in the presence of light by using water and carbon dioxide. Photosynthesis literally means "light-aided synthesis," which means that plants synthesize organic matter (carbohydrates) in the presence of light. Carbon assimilation (assimilation: absorption into the system) is another term for photosynthesis. This is represented by following equation:

$$6CO_2 + 12H_2O \xrightarrow[\text{Light}]{\text{GreenPigments}} 6O_2 + \underset{\text{carbohydrates}}{C_6H_{12}O_6}$$

During the photosynthesis process, light energy is converted into chemical energy and stored in organic matter, which is typically carbohydrate. One molecule of glucose, for example, contains approximately 686 K Calories of energy. CO_2 and water are the raw materials for this process, and oxygen and water are produced as by-products of photosynthesis. In land plants, carbon dioxide is obtained from the atmosphere through the stomata. Small quantities of carbonates are also absorbed from the soil through the roots (Kursanov et al., 1951). The starch was the visible product of photosynthesis (Sachs, 1887).

3.5.1 Site of Photosynthesis

In eukaryotes (algae and higher plants), photosynthesis occurs in cells that contain a few to many (about 1–1000) chloroplasts that vary in size and shape. Chloroplasts are distinct double membrane-bound organelles that evolved from an endosymbiotic association between free-living oxygen-evolving photosynthesis bacteria, which may have been incorporated into growing eukaryotic cells as chloroplasts. The outer chloroplast membrane is relatively permeable, whereas the inner membrane is more selectively permeable. The saclike structures known as chloroplast lamellae or thylakoids are the sites of light reactions in the chloroplast. The space within the chloroplasts is divided into two compartments, one enclosed within the thylakoids called lumen and the other outside the thylakoids called stroma. The stroma, the matrix surrounding the thylakoid, is where CO_2 is assimilated, resulting in sugar synthesis. Thylakoids can be found as grana, or unstacked and interconnected to form stroma lamellae. Each chloroplast contains 10–100 grana. Light is captured by various pigments, including chlorophyll molecules as photoreceptors for photosynthesis. These are chlorophyll-protein complexes that are involved in harvesting light energy and transporting electrons, resulting in the generation of reductant and the synthesis of ATP. The photosynthetic machinery required for light reactions exists in the plasma membrane of cyanobacteria, which forms invaginations or folded structures similar to the grana of chloroplasts in eukaryotic cells.

3.5.2 Photosynthetic Pigments

Chlorophylls, carotenoids, and phycobilins are three different types of photosynthetic pigments.

- Chlorophylls and carotenoids can only be extracted using organic solvents like acetone, petroleum ether, and alcohol because they are insoluble in water.
- Phycobillins are water-soluble.
- Carotenes and xanthophylls are examples of carotenoids. Carotenols are another name for the xanthophylls.
- Chlorophylls are the green photosynthetic pigments. Five types of chlorophylls occur in the plants other than bacteria. They are a, b, c, d, and e. Out of these, only two chlorophylls occur in the chloroplast of higher plants a and b.

3.5.3 Mechanism of Photosynthesis

In the oxidation-reduction process of photosynthesis, oxygen is released along with the oxidation of water (the removal of electrons from the water), and carbohydrates are produced as a result of the reduction of carbon dioxide. It involves two steps.

The first stage, referred to as the light reaction, is when photolysis of water takes place:

$$4e^- + 4\left[H^+\right] + CO_2 \rightarrow \left(CH_2O\right) + H_2O \quad \left(\text{Stage } II\right)$$

In the He following stage II, referred to as CO_2 assimilation, electrons extracted from water are used to reduce CO_2:

$$2H_2O \xrightarrow{\text{Light}} O_2 + 4\left[H^+\right] + 4e^- \quad (\text{Stage I})$$

As a result, light energy is transformed into chemical energy and stored as carbohydrates. While stage II is a purely chemical reaction, stage I is photochemical. The molecular mechanism underlying photosynthesis is currently fairly well understood.

Therefore, photosynthesis occurs in two phases: photochemical and biosynthetic. Photochemical phase is also called light or Hill reaction. Biosynthetic phase is also termed as dark or Blackman's reaction.

3.5.3.1 Photochemical Phase (Light or Hill Reaction)

In the light reaction, ATP and NADPH2 are created, whereas in the dark reaction, glucose is produced when CO_2 is reduced with the aid of ATP and NADPH2. Primary photochemical reaction is the name given to the light reaction because light causes it. Since Hill demonstrated that chloroplasts produce O_2 from water in the presence of light, the light reaction is also known as Hill's reaction. The reason it is also known as Arnon's cycle is that Arnon demonstrated how the H+ ions released when water breaks down are used to convert the coenzyme NADP to NADPH. As ATP is produced when light is present, photophosphorylation is a part of the light reaction. Only when light is present does the reaction occur in the grana region of the chloroplast. Since the chlorophyll absorbs light energy, it is also known as the photosystem or pigment system. Different types of chlorophyll absorb light at various wavelengths. As a result, there are two photosystems that contain chlorophyll: Photosystem I (PSI) and Photosystem II (PS II). Light with wavelengths shorter than 680 nm affects both photosystems, while light with wavelengths longer than 680 nm affects PS I.

The second phase of photosynthesis uses ATP and NADPH, which are produced during the light-dependent reactions.

Two electron transport chains produce ATP and NADPH, use water, and create oxygen during the light reactions. The photosynthesis light reaction's chemical equation can be simplified to:

$$\textbf{2H}_2\textbf{O} + \textbf{2NADP} + +\textbf{3ADP} + \textbf{3Pi} \rightarrow \textbf{O}_2 + \textbf{2NADPH} +$$
$$\textbf{3ATP}$$

3.5.3.2 Biosynthetic Phase (Dark or Blackman's Reaction)

This is the second stage of the photosynthesis process. Dark reaction refers to the chemical processes of photosynthesis that take place in the absence of light. It happens in the chloroplast stroma and is purely enzymatic and slower than the light reaction. The dark reactions also take place when light is present. In a dark reaction,

CO_2 is used to create sugars. Using the energy-rich compound ATP and the assimilatory power of the light reaction, NADPH2, the energy-poor CO_2 is converted to the energy-rich carbohydrates. Both carbon fixation and assimilation are terms used to describe the process. The reaction is also known as Blackman's reaction because Blackman proved the existence of dark reaction. Two different cyclic reactions are present in the dark reaction.

1. The C3 or Calvin cycle
2. The C4 cycle or Hatch and Slack pathway

3.5.3.2.1 The C3 Cycle or Calvin Cycle

It is a cyclic reaction that takes place during photosynthesis's dark phase. Since CO_2 is converted into sugars during this reaction, carbon fixation is taking place. Melvin Calvin was the first to notice the Calvin cycle in chlorella, a type of green algae with only one cell. For this work, Calvin received the Nobel Prize in 1961. The cycle is also known as the C3 cycle because the first stable compound in the Calvin cycle has three carbons (3-phosphoglyceric acid).

Through their stomata, plants absorb carbon dioxide from the atmosphere and start the Calvin cycle of photosynthesis.

In the Calvin cycle, six molecules of carbon dioxide are converted into one sugar molecule, or glucose, by the action of ATP and NADPH created during the light reaction (Fig. 3.3).

3.5.3.2.2 The C4 Cycle or Hatch and Slack Pathway

It is the C3 cycle's alternative pathway for fixing CO_2. In this cycle, oxaloacetic acid, a compound with four carbons, is the first stable compound to form. Thus, it is known as the C4 cycle. The pathway is also known as the C4 dicarboxylic acid pathway and is named after Hatch and Slack who developed it in 1966. Numerous grasses, sugar cane, maize, sorghum, and amaranthus all exhibit this pathway frequently (Fig. 3.4).

3.5.4 Types of Plants

3.5.4.1 C3 Plants

Plants with the C3 pathway are referred to as C3 plants. The Calvin cycle is used by these plants in the dark reaction of photosynthesis. Kranz anatomy is not visible in the leaves of C3 plants. The photosynthesis process occurs only when the stomata are open. C3 plants account for approximately 95% of all shrubs and trees, and

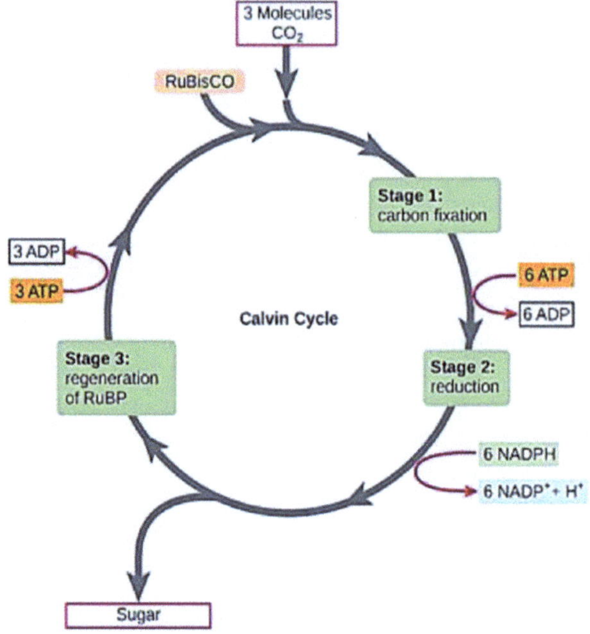

Fig. 3.3 Calvin cycle. (Kukwa & Chetty, 2020)

these plants can be either annuals or perennials. They are extremely high in protein. Wheat, oats, rye, and orchard grass are examples of C3 plants.

3.5.4.2   C4 Plants

C4 plants are those that use the C4 pathway or the Hatch-Slack pathway during the dark reaction. These plants' leaves have Kranz anatomy, and their chloroplasts are dimorphic. C4 plants account for about 5% of all plants on the planet. C4 plants are very productive in hot and dry climates, and they produce a lot of energy. C4 plants that we commonly consume include pineapple, corn, sugarcane, and so on.

3.5.4.3   CAM Plants

Entities in this type of photosynthesis absorb energy from sunlight during the day and fix carbon dioxide at night. This adaptation is seen during droughts, allowing gaseous exchange during the night when the air temperature is cooler, as well as water vapor loss. Plants such as euphorbias and cactus are examples.

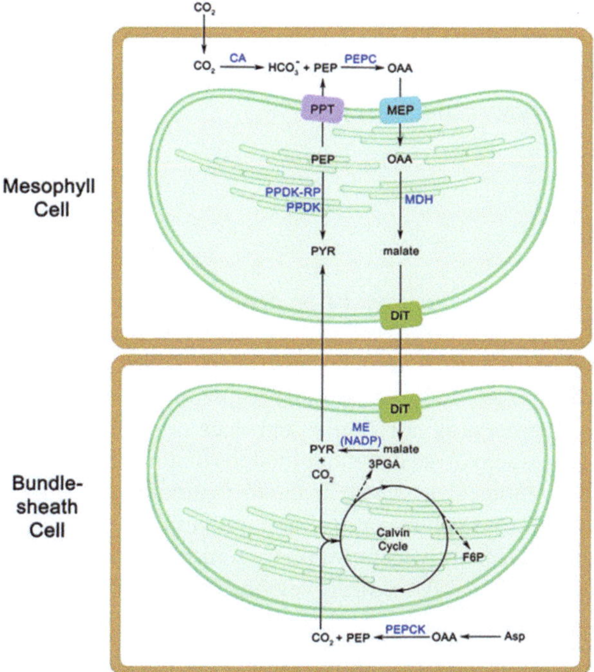

Fig. 3.4 C4 cycle. (Sheng et al., 2017)

Bromeliads and orchids have adapted to this pathway due to irregular water supply.

3.5.4.4   Efficiency of C4 Plants Over C3 Plants

1. C4 plants are more efficient than C3 plants due to their lower rate of photorespiration and higher rate of photosynthesis.
2. RuBisCO is the primary carbon fixation enzyme in the Calvin cycle.
3. It is attracted to both CO_2 and O_2.
4. When the carbon dioxide concentration is low, RuBisCO absorbs oxygen to perform photorespiration.
5. Photorespiration is greatly reduced in C4 plants due to the high carbon dioxide concentration at the RuBisCO site.
6. C4 plants have unique leaf anatomy known as "Kranz anatomy," and they use the C4 photosynthesis pathway.
7. CO_2 is first accepted by phosphoenolpyruvate (PEP) in mesophyll cells, which produces oxaloacetic acid (OAA), a 4-carbon organic acid.
8. The OAA is then converted to malic acid, which is then transported to bundle sheath cells.

9. CO_2 is released in the bundle sheath cells and enters the Calvin cycle, where it is acted on by RuBisCO.
10. Photorespiration is reduced due to the higher concentration of carbon dioxide in the bundle sheath cells.

3.6   Respiration

Respiration refers to the cellular oxidation or breakdown of carbohydrates into CO_2 and H_2O and the release of energy. It is the opposite of photosynthesis. During respiration, various organic food substances such as carbohydrates, fats, proteins, and so on may be oxidized. Glucose is the most common of these.

This oxidation process is not simple and does not occur in a single step. The breakdown of glucose involves a number of steps that result in the release of energy in the form of ATP molecules as well as the formation of a number of carbon compounds (intermediates). Respiration is a vital process that occurs in all living cells of the plant, with floral buds, vegetative buds, germinating seedlings, stem, and root apices being the most actively respiring regions.

3.6.1   Types of Respiration

3.6.1.1   Anaerobic Respiration

It occurs in the absence of oxygen, and the respiratory substrate is only partially oxidized. In addition to carbon dioxide, other compounds are formed. This type of respiration is uncommon but common in microorganisms such as yeasts.

3.6.1.2   Aerobic Respiration

In the presence of oxygen, aerobic respiration occurs, and the respiratory substrate is completely oxidized, yielding carbon dioxide and water as end products. This type of respiration is very common.

Aerobic respiration is a step-by-step catabolic process of complex oxidation of organic food into carbon dioxide and water, with oxygen acting as the terminal oxidant. It occurs via two pathways: the common pathway and the pentose phosphate pathway.

3.6.1.2.1 Common Pathway

It is known as a common pathway because its first step, glycolysis, is shared by both aerobic and anaerobic modes of respiration. The common pathway of aerobic respiration consists of three steps: glycolysis, the Krebs cycle, and terminal oxidation.

3.6.1.2.1.1 Glycolysis

It is also called EMP pathway and was discovered by three German scientists in 1930 (*Gustav Embden, Otto Meyerhof, J. Parnas*).

Glycolysis (from the Greek glykos, sweet, and lysis, splitting) is a ten-step catabolic pathway that converts a glucose molecule to two molecules of the three-carbon compound pyruvate. Two molecules of ATP are produced, and two molecules of NAD+ are reduced to NADH at the same time.

It takes place in cytoplasm or cytosol. Glycolysis occurs in both aerobic and anaerobic modes of respiration. It is the first and only step in the breakdown of glucose in aerobic respiration. Glycolysis has two stages: preparatory and payoff. During the preparatory phase, glucose is broken down to glyceraldehyde 3-phosphate. During the payoff phase, the latter is converted into pyruvate, which produces NADH and ATP.

Stage 1

- The enzyme hexokinase adds a phosphate group to glucose in the cell cytoplasm.
- A phosphate group is transferred from ATP to glucose, resulting in glucose, 6-phosphate.

Stage 2

- The enzyme phosphohexose isomerase converts glucose-6-phosphate to fructose-6-phosphate.

Stage 3

- The other ATP molecule adds a phosphate group to fructose 6-phosphate, which is then converted into fructose 1,6-bisphosphate by the enzyme phosphofructokinase.

Stage 4

- Fructose 1,6-bisphosphate is transformed by the enzyme aldolase into the isomers glyceraldehyde 3-phosphate and dihydroxyacetone phosphate.

Stage 5

- Dihydroxyacetone phosphate is transformed into glyceraldehyde 3-phosphate by triose-phosphate isomerase, which serves as the substrate for the following step of glycolysis.

Stage 6

- This step causes two reactions:
- To form NADH and H+ the enzyme glyceraldehyde 3-phosphate dehydrogenase transfers one hydrogen molecule from glyceraldehyde phosphate to nicotinamide adenine dinucleotide.
- To form 1,3-bisphosphoglycerate, glyceraldehyde 3-phosphate dehydrogenase adds a phosphate to the oxidized glyceraldehyde phosphate.

Stage 7

- Phosphate is transferred from 1,3-bisphosphoglycerate to ADP by phosphoglycerokinase to form ATP. At the end of this reaction, two molecules of phosphoglycerate and ATP are obtained.

Stage 8

- The enzyme phosphoglyceromutase moves the phosphate from the third to the second carbon of both phosphoglycerate molecules, resulting in two molecules of 2-phosphoglycerate.

Stage 9

- Enolase extracts a water molecule from 2-phosphoglycerate to produce phosphoenolpyruvate.

Stage 10

- Pyruvate kinase transfers a phosphate from phosphoenolpyruvate to ADP to form pyruvate and ATP. The end products are two molecules of pyruvate and ATP (Fig. 3.5).

Under aerobic and anaerobic conditions, pyruvate, the end product of glycolysis, is metabolized differently.

Pyruvate is metabolized via fermentation under anaerobic conditions.

Under aerobic conditions, pyruvate is oxidized in mitochondria to acetyl-CoA, which enters the citric acid cycle, and NADH from glycolysis is oxidized to NAD+ via the NADH shuttle system.

3.6.1.2.1.2 Krebs Cycle or Tricarboxylic Acid Cycle

Krebs cycle is a sequential oxidative and cyclic breakdown of activated acetate derived from pyruvate (Hans Krebs, 1930s). The cycle is also known as the tricarboxylic acid cycle, or TCA, after the initial product. It takes place in the matrix of mitochondria. It is the common pathway for the full oxidation of proteins, lipids, and carbohydrates as they are broken down into acetyl coenzyme A or other cycle intermediates. The tricarboxylic acid cycle or the citric acid cycle receives the produced acetyl CoA. In this process, glucose is fully oxidized.

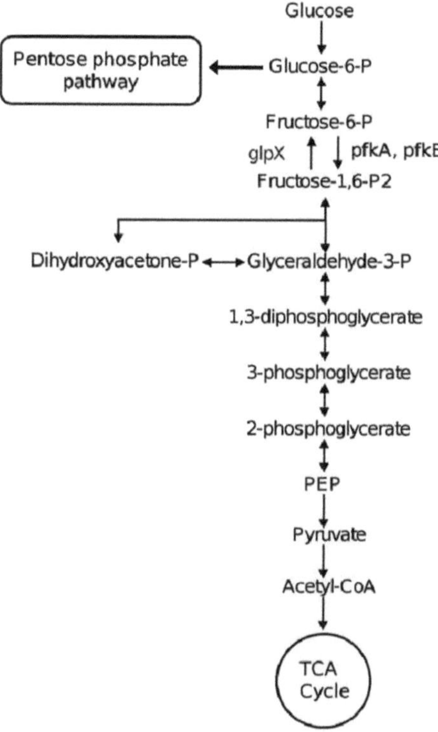

Fig. 3.5 Schematic representation of glycolytic pathway. (Phong et al., 2013)

Formation of Acetyl CoA

- Pyruvate, a product of glycolysis, enters the mitochondrial matrix. To create two molecules of Acetyl CoA, it goes through oxidative decarboxylation.

Krebs Cycle Phases

Step 1: Coenzyme A is released during the condensation of acetyl CoA with the 4-carbon compound oxaloacetate to form 6C citrate. Citrate synthase is the catalyst for the reaction.

Step 2: Isocitrate, the isomer of citrate, is produced. This reaction is catalyzed by the enzyme aconitase.

Step 3: Isocitrate is dehydrogenated and decarboxylated in step 3 to produce 5C - ketoglutarate. The release of CO2 in a molecular form. The reaction is catalyzed by isocitrate dehydrogenase. It is an enzyme that depends on NAD+. NADH is created from NAD+.

In step 4, succinyl CoA, a 4C compound, is created by the oxidative decarboxylation of -ketoglutarate. The enzyme complex known as -ketoglutarate dehydrogenase facilitates the reaction. NAD+ is changed to NADH, which results in the release of one CO2 molecule.

Step 5: Succinate is created by succinyl CoA. The reaction is catalyzed by the succinyl CoA synthetase enzyme. Along with that, GDP is phosphorylated at the

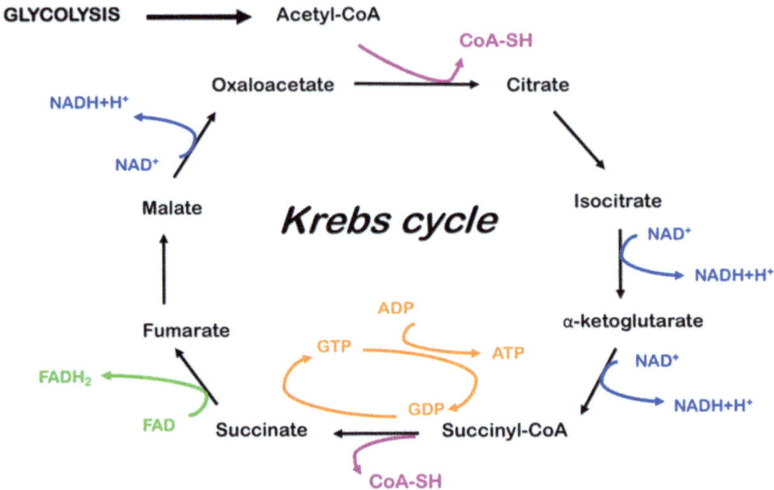

Fig. 3.6   Schematic representation of the Krebs cycle. (Protasoni & Zeviani, 2021)

substrate level to produce GTP. ATP is created when GTP transfers its phosphate to ADP.

In step 6, the enzyme succinate dehydrogenase converts succinate to fumarate. As a result, FAD is changed into $FADH_2$.

Step 7: By adding one H_2O, fumarate is transformed into malate. Fumarase is the enzyme that is catalyzing this reaction.

Step 8: Malate is dehydrogenated to create oxaloacetate, which joins forces with an additional acetyl CoA molecule to initiate a new cycle. Removed hydrogens are transferred to NAD+ to create NADH. The reaction is catalyzed by malate dehydrogenase (Fig. 3.6).

3.6.1.2.1.3   *Electron Transport System and Oxidative Phosphorylation*

When electrons from energy-rich molecules like NADH and $FADH_2$, produced in glycolysis, citric acid cycle, and fatty acid oxidation, are transferred from these molecules to molecular O_2 by a series of electron carriers, ATP is produced. O_2 is converted into H_2O. It occurs in the mitochondria's inner membrane.

The next steps is in the respiratory process that involves releasing and using the energy that has been stored in $FADH_2$ and NADH + H+. This is done by oxidizing them using the electron transport system, where the electrons are then transferred to O_2 and H_2O is produced as a result. The inner mitochondrial membrane contains the electron transport system (ETS), a metabolic pathway that allows an electron to move from one carrier to another. An NADH dehydrogenase (complex I) oxidizes the electrons from NADH produced in the mitochondrial matrix during the citric acid cycle before transferring them to ubiquinone inside the inner membrane.

Ubiquinone also receives reducing equivalents from $FADH_2$ (complex II), which is produced during the citric acid cycle's oxidation of succinate. The reduced ubiquinone (ubiquinol) is then oxidized by electron transfer to cytochrome c via the cytochrome bc1 complex (complex III). Cytochrome c is a small protein attached to the inner membrane's outer surface that acts as a mobile carrier for electron transfer between complex III and IV. Complex IV is a cytochrome c oxidase complex that includes cytochromes a and a3 as well as two copper centers.

When electrons pass from one carrier to another in the electron transport chain via complex I to IV, they are coupled to ATP synthase (complex V) for the production of ATP from ADP and inorganic phosphate. The number of ATP molecules produced is determined by the nature of the electron donor. Although the aerobic process of respiration occurs only in the presence of oxygen, the role of oxygen is limited to the process's final stage. Nonetheless, the presence of oxygen is critical because it drives the entire process by removing hydrogen from the system. The final hydrogen acceptor is oxygen.

As previously stated, the energy released during the electron transport system is used to synthesize ATP with the assistance of ATP synthase (complex V). This complex is made up of two major parts: F1 and F0. The F1 headpiece is a peripheral membrane protein complex that contains the site for ADP and inorganic phosphate

Fig. 3.7 Mitochondrial electron transport chain. (NCERT)

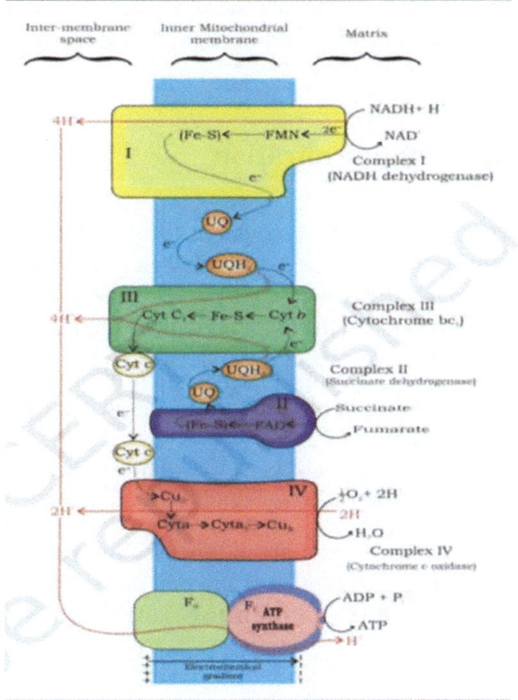

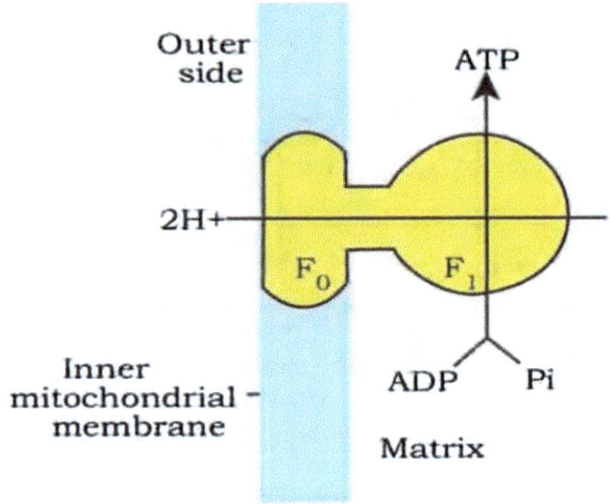

Fig. 3.8 Diagrammatic representation of ATP synthesis in mitochondria. (NCERT)

synthesis. F0 is a membrane protein complex that forms the channel that allows protons to cross the inner membrane. The passage of protons through the channel is linked to the F1 component's catalytic site for the production of ATP. 4H+ passes through F0 from the intermembranous space to the matrix down the electrochemical proton gradient for each ATP produced (Figs. 3.7 and 3.8).

3.7   Photorespiration

Dicker and Tio discovered it in 1959 (Bhatia & Tyagi, 2017). Photorespiration is the light-dependent process of ribulose biphosphate (RuBP) oxygenation and carbon dioxide release by a plant's photosynthetic organs. Normally, photosynthetic organs do not reverse in the light, taking CO_2 and releasing O_2. As a result, photorespiration is difficult to prove. It is deduced from the following:

1. A decrease in photosynthesis rate when oxygen concentration is increased from 2 to 3% to 21%.
2. A sudden increase in CO_2 evolution when an illuminated green organ is switched to darkness (Decker, 1955).

To understand more about this process, we need to understand the first step of the Calvin pathway, which is the first carbon dioxide fixation step. In this step, RuBP combines with CO_2 to form two molecules of 3PGA, which is catalyzed by RuBisCO (carboxylation).

The most abundant enzyme in the world, RuBisCO, is distinguished by the fact that its active site can bind to both CO_2 and O_2—hence the name. Which of the two will bind to the enzyme is determined by the relative concentrations of O_2 and CO_2.

As the concentration of O_2 decreases, the relative level of carboxylation increases until, at zero O_2, photorespiration is also zero. Increases in the relative level of O_2 (or decreases in CO_2), on the other hand, tip the balance in favor of oxygenation. The solubility of gases in water decreases as temperature rises, but O_2 solubility is less affected than CO_2 solubility. As a result, O_2 will inhibit photosynthesis in plants that photorespire, as measured by net CO_2 reduction. Otto Warburg discovered the inhibition of photosynthesis by O_2 in the 1920s, but it took another 50 years for the bifunctional nature of RuBisCO to provide the first satisfactory explanation for this phenomenon (Hopkins et. al., 2013). Because some O_2 binds to RuBisCO in C3 plants, CO_2 fixation is reduced. Instead of being converted to two molecules of PGA, RuBP binds with O_2 to form one molecule of phosphoglycerate and one molecule of phosphoglycolate (two carbon) in a process known as photorespiration. There is no synthesis of sugars or ATP in the photorespiratory pathway. Rather, the utilization of ATP results in the release of CO_2. There is no ATP or NADPH synthesis in the photorespiratory pathway.

Photorespiration is nonexistent in C4 plants. This is due to the fact that they have a mechanism that raises the CO_2 concentration at the enzyme site. This occurs when the bundle sheath cells break down the C4 acid from the mesophyll to release CO_2, which raises the intracellular concentration of CO_2. In turn, this makes sure that the RuBisCO minimizes oxygenase activity while acting as a carboxylase.

3.8   Evapotranspiration

Evapotranspiration (ET) is the combination of two distinct processes whereby water is lost from the crop by transpiration and from the soil surface by evaporation, respectively. Evapotranspiration process involves transpiration and evaporation.

3.8.1   Evaporation

The process of converting liquid water into water vapor (vaporization) and removing it from the evaporating surface is called evaporation (vapor removal). Many different surfaces, including lakes, rivers, pavements, soils, and wet vegetation, allow water to evaporate.

The transition of water molecules from their liquid state to their gaseous state requires energy. This energy is provided by direct solar radiation and, to a lesser extent, by air temperature. The difference between the water vapor pressure at the evaporating surface and that of the surrounding atmosphere is what propels the removal of water vapor from the evaporating surface. If the wet air is not transferred to the atmosphere, evaporation will slow down and possibly stop as the surrounding air gradually becomes saturated. The amount of wind that blows determines how quickly the saturated air is replaced by dryer air. Therefore, when evaluating the evaporation process, climatological parameters such as solar radiation, air temperature, air humidity, and wind speed should be taken into account.

3.8.2   Transpiration

Transpiration is the term used to describe the loss of water from a plant's exposed areas as vapor. As much as 1 ton of water per day can be lost through transpiration in elm trees, 36–45 L per day in apples, and 2 L per day in sunflowers. Instead, transpiration causes 98–99% of the water that a plant takes in to be lost. Only 0.2% is used in photosynthesis, while the remaining is retained in the plant during growth. Transpiration mainly occurs from the foliar surface of the plant, and it accounts for over 90% of the total transpiration. Transpiration also occurs from young stems flowers fruits (Bhatia & Tyagi, 2017).

Similar to direct evaporation, transpiration is influenced by the energy source, vapor pressure gradient, and wind. Therefore, when evaluating transpiration, radiation, air temperature, air humidity, and wind terms should be taken into account. Different plant species may transpire at varying rates. When evaluating transpiration, it is important to take into account not only the type of crop but also its development, environment, and management.

3.8.3   Evapotranspiration (ET)

Evapotranspiration (ET) is the collective name for the processes by which water is lost to the atmosphere from the earth's surface. It includes both water transpiration and water evaporation (the direct transfer of water from soil, canopies, and water bodies to the air) (movement of water from the soil, through roots and bodies of vegetation, on leaves and then into the air). Measurement of evapotranspiration is crucial for managing water resources and irrigating agriculture because it is a crucial component of the local water cycle and climate.

3.8.4 Units

The evapotranspiration rate is typically expressed in millimeters (mm) per unit time. The rate expresses the amount of water lost from a cropped surface in units of water depth. The time unit can be an hour, day, decade, month, or even an entire growing season or year.

3.8.5 Factors Affecting Evapotranspiration

Evapotranspiration is affected by a variety of factors, including the following:

- Temperature: As the temperature rises, so does the rate of evapotranspiration. Evaporation accelerates because there is more energy available to convert liquid water to water vapor. Transpiration increases as temperatures rise, because plants open their stomata and release more water vapor.
- Humidity: If the air around the plant is too humid, transpiration and evaporation rates decrease.
- Wind speed: As the air moves, the rate of evaporation increases. The wind will also clear the air of any humidity produced by the plant's transpiration, causing the plant to increase its rate of transpiration.
- Water availability: If the soil is dry and there is no standing water, there will be no evaporation. If plants cannot get enough water, they will conserve it by closing their stoma rather than transpiring
- Soil type: It determines how much water a soil can hold and how easily water can be drawn out of it, either by a plant or by evaporation. The rate of transpiration is significantly higher than the rate of evaporation from the soil in areas where the ground is covered by vegetation.
- Plant type: Some plants, such as cacti and succulents, naturally retain water and do not transpire as much. Trees and crops, on the other hand, can emit massive amounts of water vapor in a single day.

3.9 Plant Growth Regulators

Plant growth regulators are small simple molecules with diverse chemical compositions that regulate growth differentiation and development at low concentrations by promoting or inhibiting the same. Plant hormones, their synthetic analogues, inhibitors of hormone biosynthesis, and blockers of hormone receptors are all examples of PGRs. The term "hormone" was first used in medicine about 100 years ago to describe a stimulatory factor. It is a Greek word meaning "to stimulate" or "to set in motion." According to the classic definition of a plant hormone, it is a substance produced in any part of the organism and transferred to another part to influence a specific physiological process (Went & Thimann, 1937).

There are five different kinds of plant hormones or phytohormones. They are adenine derivatives (such as furfuryl, aminopurine, kinetin, and cytokinins), indole compounds (such as indole acetic acid or IAA), derivatives of carotenoids and fatty acids (such as abscisic acid ABA), terpenes (such as gibberellins, such as GA3), and gases (ethylene). Brassinosteroids, jasmonic acid, and salicylic acid are additional related growth regulators. Some vitamins also act as PGRs (*Elementary biology by K N Bhatia and M P Tyagi*).

3.9.1   Auxins

Auxins are a class of naturally occurring hormones that have been found in virtually all land plants as well as several soil- or plant-associated microbes. Indole-3-acetic acid (IAA), indole-3-butyric acid (IBA), phenyl acetic acid, and 4-chloroindole-3-acetic acid (4-Cl-IAA) are all naturally occurring auxins. IAA is the most extensively researched auxin. The presence of an acidic side chain on the aromatic ring is a chemical feature shared by all molecules exhibiting auxin activity. Except for phenyl acetic acid, all of the natural auxins listed above are indole derivatives. One of the first major plant hormones discovered was auxin. In 1928, Dutch botanist Fritz Went isolated auxin from the tip of the oat coleoptile in a gelatin block for the first time.

3.9.1.1   Functions of Auxins

- Respiration: Auxins most likely stimulate respiration by increasing the availability of respiratory substrate.
- Metabolism: The use of auxins has been shown to improve metabolism by mobilizing plant resources.
- Cell enlargement: The most fundamental activity of auxins is cell enlargement. Solubilization of carbohydrate, loosening of wall microfibrals, synthesis of more wall materials, increased membrane permeability, and respiration all contribute to cell enlargement.
- Cambial activity: The degree of cambial activity is proportional to the concentration of auxin (Avery et al., 1947). Auxin has been shown to regulate xylem differentiation.
- Cell division: Auxin is known to promote cell division in vascular cambium cells.
- Tissue culture: It promotes the formation of a callus or a mass of undifferentiated cells in tissue culture.
- Apical dominance: Apical dominance is the phenomenon in which the presence of an apical bud prevents nearby lateral buds from growing. When the apical bud

is removed, the lateral buds sprout completely, resulting in dense bushy growth. By releasing auxins, the apical bud inhibits the growth of lateral buds.

- Sex: Auxins have a feminizing effect on some plants.

3.9.2 Cytokinins

Cytokinins (CK) are a type of plant growth hormone that promotes cell division. Miller et al. discovered the first cytokinin from Herring (an oily fish from the genus Clupea) sperm DNA in 1955. Skoog and his colleagues tested many substances in the 1940s and 1950s for their ability to initiate and sustain proliferation of cultured tobacco pith tissue. When cultured pith tissue was treated with autoclaved Herring sperm DNA, they observed cell division stimulation. This showed that the DNA degradation product stimulated cell division in tobacco pith culture. Because it caused cytokinesis, this compound was named kinetin. It is now known as 6-furfurylaminopurine. Kinetin is a naturally occurring compound, but it is not synthesized in plants. As a result, it is classified as a "synthetic cytokinin" in plants. Following that, it was discovered that immature endosperm from corn (*Zea mays*) contained a substance with biological activity similar to kinetin. When combined with auxin, this substance stimulates the division of mature plant cells. Zeatin [trans-6-(4-hydroxy-3-methyl-2-butenylamino) purine] was later identified as the active ingredient. Miller and Letham discovered zeatin, the first natural cytokinin, in unripe maize kernels in 1963.

A large number of synthetic compounds have been created and tested for cytokininactivity. Benzylaminopurine (BAP), N,N'diphenylurea, thidiazuron (TDZ), and benzyladenine are a few examples.

3.9.2.1 Functions of Cytokinins

- Cell division: Cytokinins are required for cytokinesis, but chromosome doubling can occur without them. Cytokinins cause cell division in the presence of auxins, even in permanent cells. Both hormones are required for cell division in callus.
- Cell elongation: Cytokinins, like auxins and gibberellins, cause cell elongation.
- Morphogenesis: Auxins and cytokinins are both required for tissue and organ differentiation. When cytokinin levels are high, buds form, while roots form when the ratios are reversed (Skoog and Miller, 1957).
- Differentiation: Cytokinins stimulated the formation of new leaves, chloroplasts in leaves, lateral shoot formation, and the formation of adventitious roots. They also cause lignification and differentiation of the interfascicular cambium.
- Apical dominance: The presence of cytokinins in an area causes nutrients to move toward it. They aid in the growth of lateral buds despite the presence of an apical bud. As a result, they act in opposition to auxin, which promotes apical dominance.

- Flowering: In some cases, cytokinins can replace the photoperiodic requirement for flowering.
- Parthenocarpy: Parthenocarpy has been reported to be induced by cytokinin treatment (Crane, 1956).

3.9.3 Gibberellins

Gibberellins are growth hormones that promote cell elongation and have an impact on a variety of developmental processes such as stem elongation, seed germination, dormancy, flowering, sex expression, enzyme induction, and leaf and fruit senescence. Japanese researchers discovered a common disease that causes excessive growth in rice plants. Kurosawa (1926) investigated this bakanae (foolish seedling) disease in rice and discovered that the fungus that infected the plants secreted a chemical that caused the diseased rice plants to grow tall. This chemical was isolated from the cultured fungus' filtrates and named gibberellin after the fungus that infects rice plants, Gibberellafujikuroi (now renamed Fusariumfujikuroi). Yabuta and Hayashi crystallized the fungal growth-inducing factor gibberellin from the fungus Gibberellafujikuroi in 1935. Technically, all gibberellins are diterpene acids. Plants contain a variety of gibberellins, only a few of which are biologically active as hormones. Gibberellins with 19 carbons are generally biologically active. GA1, GA3, and GA4 are the three most common biologically active gibberellins.

3.9.3.1 Functions of Gibberellins

- Germination of seeds: Some light-sensitive seeds, such as tobacco and lettuce, germinate poorly in the absence of sunlight. If the seeds are exposed to sunlight, they germinate quickly. The light requirement can be met if the seeds are treated with gibberellic acid.
- Buds dormancy: Autumn buds remain dormant until the following spring. This dormancy can be broken by administering gibberellin.
- Root development: Gibberellins have almost no effect on root growth. However, some growth inhibition can occur at higher concentrations in a few plants.
- Promotes bolting: This hormone promotes bolting, or internode elongation, in beets, cabbage, and many other plants with a rosette habit.
- Delay senescence: Senescence is postponed by gibberellins. As a result, the fruits can be left on the tree for an extended period of time to extend the market period.
- Vernalization: Application of gibberellins places the vernalization or low temperature requirement of some plants.
- Fruit development: Gibberellins control fruit growth and development along with Auxins.
- Flowering hormone: This hormone promotes flowering in long-day plants.

- Parthenocarpy induction: Gibberellins can help induce parthenocarpy development of seedless fruits from unfertilized pistils especially in case of pomes example apple and pear.

3.9.4 Ethylene

Ethylene is a plant growth regulator that is widely used for fruit ripening and the production of more flowers and fruits. Cousins discovered in 1910 that ripe oranges produce a volatile substance that accelerates the ripening of nearby unripe bananas. R. Gane (1934) discovered that the ripening-causing volatile substance was ethylene using gas chromatography. Crocker et al. (1935) identified ethylene as a plant hormone. Plants produce ethylene from the amino acid methionine. It can be found in almost all plant parts, including roots, leaves, flowers, fruits, and seeds (Denny & Miller, 1935).

3.9.4.1 Functions of Ethylene

- Senescence: It is known to hasten the senescence of leaves and flowers.
- Abscission: Ethylene stimulates the abscission of various parts, such as leaves, flowers, and fruits, and this causes the formation of hydrolases.
- Apical dominance: Apical dominance is encouraged by ethylene, and lateral buds' dormancy is prolonged.
- Breaking of dormancy: It breaks the dormancy of seeds, buds, and storage organs.
- Root initiation: Ethylene supports the growth of lateral roots, root hairs, and the initiation of new roots in low concentrations. Plant roots now have a larger surface area for absorption.
- Fruit ripening: It promotes the ripening of climacteric fruits and the dehiscence of dry fruits. Climacteric fruits are fleshy fruits that exhibit a sudden sharp increase in respiration rate during ripening. They are usually transported in the green or unripe stage. Ethylene is used in the artificial ripening of those fruits, for instance, apple, mango, and banana.
- Flowering: It promotes flowering in pineapple and related plants, as well as mango, but in some cases, the gaseous hormone causes flower fading. This aids in the synchronization of the fruit set.

3.9.5 Abscisic Acid

Abscisic acid is an inhibitory hormone that aids plant stress adaptation. It also keeps the water balance balanced, prevents seed embryos from germinating, and induces seed and bud dormancy. It is also referred to as dormin because it causes dormancy in buds, underground stems, and seeds. In 1963, Fredrick T. Addicott and his colleagues attempted to identify abscisic acid in cotton fruits. It is found in many parts of plants, but it is most abundant in green cells' chloroplasts. Mevalonic acid or xanthophylls form this hormone. It is transported to all parts of the plant via diffusion as well as phloem and xylem transport channels.

3.9.5.1 Functions of Abscisic Acid

- Bud dormancy: Abscisic acid induces dormancy in buds as winter approaches.
- Seed dormancy: Abscisic acid is the primary cause of seed dormancy. Dormancy allows seeds to withstand desiccation and temperature extremes better, only when gibberellins overcome abscisic acid do buds and seeds sprout. It is also known as dormin because of its ability to induce dormancy.
- Leaf senescence: Its excessive presence inhibits protein and RNA synthesis in the leaves, hastening their senescence.
- Transpiration: ABA is rapidly synthesized during desiccation and other stresses. The inhibitor causes stomata to close, preventing transpiration.
- Resistance: Abscisic acid boosts plant resistance to cold and other stresses. As a result, it is also known as stress hormone.

3.10 Transpiration

Transpiration is the term used to describe the loss of water from a plant's exposed areas as vapor. As much as 1 ton of water per day can be lost through transpiration in elm trees, 36–45 L per day in apples, and 2 L per day in sunflowers. Instead, transpiration causes 98–99% of the water that a plant takes in to be lost. Only 0.2% is used in photosynthesis, while the remaining is retained in the plant during growth. Transpiration mainly occurs from the foliar surface of the plant, and it accounts for over 90% of the total transpiration. Transpiration also occurs from young stems flowers fruits (Bhatia & Tyagi, 2017).

3.10.1 Types of Transpiration

Depending upon the surface of the plant, transpiration is of following four types: cuticular, lenticular, bark, and stomatal.

1. Cuticular: Transpiration that occurs through the cuticle or epidermal cells of leaves and other expose parts of plant is called cuticular transpiration. It accounts for about 3–10% of the total transpiration in common land plants. But in case of

herbaceous shade loving plants, the cuticular transpiration maybe up to 50% of the total due to their very thin cuticle.

2. Lenticular transpiration: It accounts for only 0.1% of total transportation and is found only in the woody branches of the trees where lenticels occur. It occurs during the day as well as the night because lenticels have no mechanism of closure.
3. Bark transpiration: This transpiration occurs through corky covering of the stems. It is very little, but due to the larger area, its measured rate is often more than lenticular transpiration. Like cuticular and lenticular transpiration, it occurs during the day and night.
4. Stomatal transpiration: This type of transpiration occurs through the stomata. The stomata are found mostly on the leaves. They may also occur on young stems, flowers, and fruits. It accounts for about 50–97% of the total transpiration. And continuous till the stomata are kept open. Stomatal transpiration is the most important type of transpiration.

The following section discusses the stomatal transpiration in detail.

3.10.2 Structure of the Stomata

The tiny openings present on the epidermis of leaves are called stomata. They play an important role in gaseous exchange and photosynthesis.

Stomata are made up of minute pores called stoma that are surrounded by a pair of guard cells. Stomata are open and close according to the turgidity of guard cells. The pore is surrounded by a tough and flexible cell wall. The shape of guard cells usually differs in both monocots and dicots, though the mechanism continues to be the same. Guard cells have chloroplasts and are bean-shaped. They have chlorophyll and absorb light energy.

The guard cells are surrounded by subsidiary cells. They are accessory cells to guard cells that are found in plant epidermis. They are found between guard cells and epidermal cells and serve to protect epidermal cells when guard cells expand during stomatal opening.

The average number of stomata per square mm of leaf surface is around 300.

3.10.2.1 Types of Stomata

Stomata were classified on the basis of number and arrangement of subsidiary cells (Metcalfe & Chalk, 1950).

1. Anomocytic stomata: The stomata remains surrounded by a limited number of subsidiary cells, and the subsidiary cells are quite alike of the other epidermal cells, for example, *Papaveraceae.*

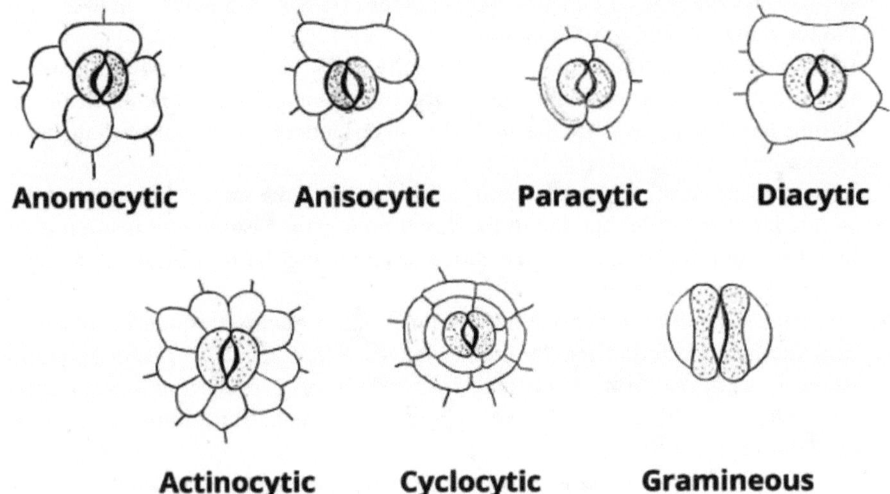

Fig. 3.9 Types of stomata

2. Anisocytic stomata: The stomata is surrounded by three subsidiary cells. One of the three cells is distinctly smaller than the other two, for example, *Cruciferae* and *Solanum*
3. Paracytic stomata: The stomata is surrounded by two subsidiary cells, and they are parallel to the longitudinal exes of the pore and guard cells.
4. Cyclocytic stomata: Four or more subsidiary cells surrounding the stoma in a narrow ring are present, for example, *Pandanus.*
5. Diacytic stomata: A pair of subsidiary cells surrounds the stomata and their common wall is at right angles to the guard cells, for example, *Caryophyllaceae.*
6. Actinocytic stomata: Four or more subsidiary cells that elongate radially to the stomata are present in this type of stomata, for example, *Araceae.*
7. Graminaceous type: Each stoma possesses to dumbbell-shaped guard cells, and the subsidiary cells are parallel to the guard cells full stop, for example, *Poaceae* (Fig. 3.9).

3.10.2.2 Mechanism of Stomatal Transpiration

There are three steps that can be taken to study the mechanism of stomatal transpiration that occurs during the day.

1. The mesophyll cells in the leaf allow water to smear from the xylem to the intercellular space above the stomata through osmotic diffusion.
2. The stomata's opening and closing (stomatal movement).
3. Simple diffusion of water vapor through stomata from intercellular spaces to other atmosphere.

- The xylem and intercellular space above the stomata are in contact with the mesophyll cells inside the leaf. Mesophyll cells become turgid when they draw water from the xylem, and as a result, their osmotic pressure (OP) and diffusion pressure deficit (DPD) both decrease. This causes them to release water in the form of vapor in intercellular spaces close to stomata through osmotic diffusion. Now that the OP and DPD of the mesophyll cells have increased, they can draw water from the xylem through osmotic diffusion.
- Opening and closing of stomata: Following an increase in the guard cells' osmotic pressure (OP) and diffusion pressure deficit (DPD) (caused by an accumulation of osmotically active substances), water is osmotically diffused into the guard cells from the mesophyll and epidermal cells around them. The guard cells' turgor pressure (TP) rises as a result, making them turgid. The guard cells enlarge, lengthen, and starch form pores on their adjacent thickened surfaces, allowing the stomata to open.

  On the other hand, water is released back into the surrounding epidermal and mesophyll cells by osmotic diffusion, and the guard cells become flaccid when OP and DPD of guard cells decrease relative to those of the surrounding epidermal and mesophyll cells (due to depletion of osmotically active substances). The guard cells' thickened surfaces come together, closing the stomatal pore and stomata as they do so.
- The process of water vapor simply diffusing from the intercellular spaces to the atmosphere through open stomata is the final stage of transpiration. This is due to the intercellular spaces being more saturated with moisture than the surrounding stomata in comparison to the outer atmosphere.

3.10.3 Significance of Transpiration

1. Water and minerals are transported to various parts of the plants by transpiration.
2. A balance of water is kept within the plant because water is continuously removed from the plant's body.
3. Osmosis is maintained, and the cells are kept rigid.
4. Transpiration generates a suction force that aids in the upward movement of water in plants.
5. The leaves are kept moist by the accumulation of some hydrophilic salts on their surface.
6. It promotes cell division and keeps the turgidity of the cells.
7. The proper growth of the plants is supported by optimal transpiration.
8. The evaporation of water from a tree's leaves is what gives it its cooling effect.

3.10.4   Guttation

Guttation is the loss or excretion of water from the leaves and other parts of an undamaged or intact plant in the form of liquid droplets (Studied by Bergerstein, 1887). Not all plants exhibit this process. It is limited to about 345 genera of herbaceous marshy and mesophytic plants, as well as some woody ones. Oats, tomatoes, cucurbits, and other common foods are examples.

It occurs through unique structures known as hydathodes. They are typically located on the edges and tips of leaves.

Numerous chemical and inorganic elements, including carbohydrates and minerals like potassium, may be present in guttation fluid. If the guttation fluid has a high nitrogen content, it may occasionally be a sign of particular events, such as fertilizer burns.

3.10.5   Stem Bleeding

It is the sap or watery solution that the plant's cut or injured parts exude, such as in *Agave* and *Vitis*. Root pressure, phloem pressure, local xylem pressure, and latex or resins all contribute to its occurrence.

MCQs

1. Plant cells are typically between ___ and ___ micrometres in length

   - 100 and 1000
   - 50 and 100
   - 10 and 100
   - 1 and 10

2. How much water does a seed on an average contain?

   - 10–15 %
   - 50–60 %
   - 20–30 %
   - 15–25 %

3. The process by which water enters the cell sap of a living plant cell by osmosis when the plant cell is immersed in water or a hypotonic solution with an O.P. lower than the cell sap is called?

   - Imbibition
   - Endoosmosis

- Exoosmosis
- Guttation

4. When a plant cell or tissue is immersed in a hypertonic solution, water expels from the cell sap into the surrounding solution by the process of?

- Endoosmosis
- Guttation
- Exoosmosis
- None of the above

5. Criteria for essentiality of elements were given by?

- Reddy and Reddy
- Hatch and Slack
- Watson and Crick
- Arnon and Stout

6. How many essential elements are there?

- 17
- 26
- 50
- 7

7. Which on of the following is classified as a benefitial element

- Selenium
- Carbon
- Magnesium
- Hydrogen

8. One molecule of glucose, for example, contains approximately _____ of energy

- 686 kCal
- 485kCal
- 984 kCal
- 527 kCal

9. Kranz Anatomy is seen in?

- CAM Plants
- C3 Plants
- C4 Plants
- All of the above

10. Which of the PGR's help in ripening of the fruit?

- ABA
- Ethylene
- Auxins
- Cytokinins

References

Armstrong, D. (2001). *A biographical dictionary of twentieth-century botanists*. The University of Chicago Press.

Avery, G. S., Burkholder, P. R., & Creighton, H. B. (1947). Hormonally controlled differentiation in the embryos of corn induced by gibberellic acid. *Proceedings of the National Academy of Sciences, 33*(4), 92–101.

Bhatia, K. N., & Tyagi, M. P. (2017). *Elementary Biology* 4; 12, 24, 31, 32, 33 43, 44, 90, 102, 107, 130.

Class 11th NCERT., Biology, *13,* 220.

Crane, J. C. (1956). Growth substances in fruit setting and development. *Annual Review of Plant Physiology, 7*(1), 383–406.

Crocker, W., Hitchcock, A. E., & Zimmerman, P. W. (1935). Similarities in the effects of ethylene and the plant auxins. *Contributions from Boyce Thompson Institute, 7,* 231–248.

Decker, J. P. (1955). A comparative study of the respiration of wild and cultivated plants. *Ecology, 36*(1), 1–14.

Denny, F. E., & Miller, E. V. (1935). The behavior of ethylene in the ripening of bananas. *Contributions from Boyce Thompson Institute, 7,* 97–102.

Gane, R. (1934). Production of ethylene by some ripening fruits. *Nature, 134*(3394), 1008.

Hopkins, W. G., et al. (2009). *Introduction to plant physiology* (4th ed., pp. 62–76, 144). John Wiley & Sons.

Hopkins, W. G., Huner, N. P. A., & Shipley, B. (2013). *Introduction to plant physiology*. Wiley.

https://legacy.climate.ncsu.edu/edu/Evap

https://www.biologydiscussion.com/plants/absorption-of-water/absorption-of-water-in-plants-with-diagram/22718

Kukwa, D., & Chetty, M. (2020). *Microalgae: The multifaceted biomass of the 21st century.* https://doi.org/10.5772/intechopen.94090

Kursanov, A. L., et al. (1951). Movement of assimilates. Trudy Instituta Fiziologii Rastenii Imeni K.A. *Timiryazeva, 14,* 5–93.

Malik, C. P., & Srivastava, H. N. (1982). *Textbook of plant physiology*. Kalyani Publishers.

Metcalfe, C. R., & Chalk, L. (1950). *Anatomy of the Dicotyledons*. Clarendon Press.

Meyer, A. (1938). *Plant physiology* (Vol. 2). McGraw-Hill.

Noggle, G. R., & Fritz, G. J. (1983). *Introductory plant physiology*. Prentice-Hall.

Phong, W., Lin, W., Rao, S., Dick, T., Alonso, S., & Pethe, K. (2013). Correction: Characterization of phosphofructokinase activity in mycobacterium tuberculosis reveals that a functional glycolytic carbon flow is necessary to limit the accumulation of toxic metabolic intermediates under hypoxia. *PloS One, 8,* e56037. https://doi.org/10.1371/journal.pone.0056037

Protasoni, M., & Zeviani, M. (2021). Mitochondrial structure and bioenergetics in normal and disease conditions. *International Journal of Molecular Sciences., 22,* 586. https://doi.org/10.3390/ijms22020586

Renner, O. (1912). Untersuchungen über die Reizbewegungen der Pflanzen. *Zeitschrift für Botanik, 4,* 65–96.

Sachs, J. (1887). *Lectures on the physiology of plants*. Clarendon Press.

Sardare, M. (2013). A review on plant without soil – Hydroponics. *International Journal of Research in Engineering and Technology., 2,* 299–304.

Sheng, J., Zheng, X., Wang, J., Zeng, X., Zhou, F., Jin, S., Hu, Z., & Diao, Y. (2017). Transcriptomics and proteomics reveal genetic and biological basis of superior biomass crop Miscanthus. *Scientific Reports, 7.* https://doi.org/10.1038/s41598-017-14151-z

Skoog, F., & Miller, C. O. (1957). Chemical regulation of growth and organ formation in plant tissues cultured in vitro. *Symposia of the Society for Experimental Biology, 11,* 118–130.

Slatyer, R. O., & Taylor, S. A. (1960). Terminology in plant-and soil-water relations. *Nature, 187*(4744), 922–924.

Went, F. W., & Thimann, K. V. (1937). *Phytohormones*. Macmillan.

Chapter 4
Crop Production Technology

Abstract This chapter covers all the production practices of field crops. It has been carefully drafted, with utmost care regarding the figures and data used, and it's made sure they're nationally and internationally recognized. Crop production depends on the successful execution of land, water, and nutrient management practices from the time of sowing to the time of harvesting. This chapter introduces readers to the concept of Kharif, Rabi, and Zaid crops along with the production practices of these crops under various headings and subheadings.

Keywords Origin · Kharif · Rabi · Zaid · Fertilizers · Nutrients · Variety · Cultivar · Intercultural operations · Yield

4.1 Introduction

In this chapter, we will be studying about the production practices of all the major field crops. The field crops have been divided according to the seasons, that is, Kharif season, Rabi season, and Zaid season.

First, we'll take a brief look at the classification of crops based on seasons, and after that, we'll study in detail the production technology of each crop.

1. *Kharif Crops:* Kharif crops, monsoon crops, or autumn crops are cultivated and harvested in the monsoon season. The farmer sows seeds at the beginning of the monsoon season and harvests them at the end of the season, that is, between September and October. Kharif crops need a lot of water and hot weather for proper growth. Some of the major Kharif crops are as follows:

   - Rice, maize, bajra, and jowar (*cereals*)
   - Cotton and jute (*fiber crops*)
   - Groundnut and soybean (*legumes*)
   - Sugarcane (*sugar crops*)
   - Moong (green gram) (*pulses*)

2. *Rabi Crops:* Arabic for "rabi" is "spring." Rabi crops are those that are cultivated during the winter months (from October to December) and harvested during the spring months (from April to May). These crops need a warm climate for seed germination and seed maturity and a cold environment for growth. Winter rain damages the Rabi crop but benefits the Kharif crop. Some of the major Rabi crops are as follows:

   - Wheat and barley (*cereal*)
   - Pea and lentils (*pulses*)
   - Oats (*forage*)
   - Linseed, sunflower, and mustard (*oilseed*)

3. *Zaid Crops:* Zaid crops are planted from March to June, in the period between the Kharif and Rabi seasons. They need longer days for flowering and warm and dry weather for their critical growth period. Zaid crop is important for farmers because it provides them with quick income and fills the gap between their two main crops, Kharif and Rabi. Some of the major Rabi crops are as follows:

   - Cowpea and black gram (*pulses*)

4.2 Production Practices

This section deals with the production practices of the major field crops.

4.2.1 Production Practices of Kharif Crops

Table 4.1 shows the production practices of cereal crops.
Table 4.2 shows the production practices of fiber crops.
Table 4.3 shows the production practices of legumes.
Table 4.4 shows the production practices of sugar crops.
Table 4.5 shows the production practices of pulses.

4.2.2 Production Practices of Rabi Crops

Table 4.6 shows the production practices of rabi cereal crops: wheat and barley.
Table 4.7 shows the production practices of Rabi pulses: pea and lentil.
Table 4.8 shows the production practices of Rabi forage crop: oats.
Table 4.9 shows the production practices of Rabi oilseeds: linseed, mustard, and sunflower.

Table 4.1 Production practices of Kharif cereal crops

S.No	Cereal	Rice	Maize	Jowar	Bajra
1.	Botanical name Chromosome no.	*Oryza sativa* $2n = 24$	*Zea mays* $2n = 20$	*Sorghum bicolor* $2n = 20$	*Pennisetum americanum* $2n=20$
2.	Soil type	Acidic saline alkali, clay, silt clay/silty clay loam, marshy soils loamy, pH 4.5–8.5	pH ranges from 5.5 to 7.5; water logging for 3 days reduces the yield by 40–45% well drained sandy loam to silt loam	Grown on deep, well drained permeable soils pH range of 6.0–8.5, sensitive to aluminum toxicity and soils with acid saturation	Low fertility, mid saline, black, alluvial and red sandy, pH 5.5–8
3.	Climate	High humidity, prolonged sunshine rain falls 800–1000 mm, 20–25 °C	12–15 °C during growing period and around 30 °C during maturity	It can tolerate high temperature throughout its life cycle, 26–30 °C temperature for good growth. Annual rainfall between 45 and 65 cm	Warm weather, 400–750 mm, tolerate frost, 30–35 °C
4.	Land preparation	Ploughing, harrowing, planking, and tilth must be good and firm for better germination. Puddling: Ploughing land with standing water until the soil becomes soft and muddy	Prepared flat-beds which has given 4–5 deep ploughing provided an ideal condition	Three systems of sorghum sowing are followed: Sowing on a flat surface, or ride and furrow, broad bed and furrow	Land should be made weed free from previous planting and land ploughed six to seven time to achieve fine tilth stage
5.	Sowing time	Kharif: from June 15 to July 7	Kharif: month of February/March. Rabi: middle of October Oct to middle of November	Kharif- June-July Rabi Mid Sept-Mid Oct Summer- Mid Jan-Mid Feb	From June 15 to July 15
6.	Seed rate	Broadcasting:100 kg Transplanting: 60 kg *Hybrid* 15 kg. Dapog 2 3 kg/m^2 (4–6 seedling/hill) @ 14 DAS *SRI*: 5–6 Kg/ha @ 12 DAS	Composite: 20 kg/ha, hybrids: 5 kg/ha	Irrigated: Transplanted 7,5 kg/ha; direct sown 10 kg/ha rainfed (direct sown) 15 kg/ha	3–5 kg/ha

(continued)

Table 4.1 (continued)

S.No	Cereal	Rice	Maize	Jowar	Bajra
7.	Seed treatment	Dissolve 3 g fungicide (e.g., Benlate + Mancozeb or Arazone red alone) per kg seed in 5 ml water Inside a plastic bag	Bavistin @ 3 gm/Kg of seeds before sowing in the field	300 mesh sulfur powder @ 4 g of sulfur per kg of seeds for controlling the smut disease. Seed soaked in 30% salt solution	Thiram or cabendazim @ 2: 2.5 gm per 1 kg of seed before sowing to avoid seed-borne diseases
8.	Spacing	20 × 10 cm 25 × 25 cm 30 × 30 cm	60 × 20 cm, 83,333 plants/ha. For maximum yield: 1.1 l/ha with (60 × 15 cm)	40–50 cm × 10–15 cm	40 × 10 cm
9.	Fertilizer Management	FYM 5 to 10 t/ha; Azolla 6 t/ha, 2–3 t/ha lime (15–30 days before sowing) *Kharif:* 100:60:60 kg/ha *Summer:* 125: 62.5:62.5 kg/ha *basal dose* 50%N full P&K Tillering (30DAS) 25%, panicle initiation 25%	12.5 t FYM/ha *IR:* 150:75:40 kg/ha *RF:* 100:50:25 kg/ha $ZnSO_4$-10 kg/ha for whiptip	10–15 t/ha of FYM 4 5 weeks before sowing *IR:* 100:75:40 kg/ha *RF:* 65:40:40 kg/ha	10–15 FYM/ ha *IR:* 100:65:65 kg/ha *RF:* 50:25:10 kg/ha, foliar spray of 2–3% urea
10.	Water management	800–2500 mm/ha 1 kg rice = 500 L seedling, transplanting, tillering, panicle, booting, milky dough stage	Requires 500–600 mm Critical stages for irrigation are tasselling and silking	Critical stage of irrigation - flowering, grain filling	Not in kharif but req @ critical growth stages, 45–50 ha cm, tillering, panide, soft /hard dough
11.	Weed management	Use Butachlor 1.25 kg/ha or Anilophos 0.4 kg/ha as pre-emergence application 2,4 D, sodium salt post emergence	Apply herbicides Atrazine @ 1–1.25 kg a.e/ha (pre- emergence) 2, 4 D @ 0.75 kg a.e/ha 30–35 DAS	Apply herbicides Atrazine @ 1–1.25 kg a.e/ha (pre- emergence) 2, 4 D @ 0.75 kg a.e/ha 30–35 DAS	Use Atrazine @ 1.25 kg/ha. Pendimethalin 30EC @ 2.5 L with atrazine to control weeds

12.	Harvesting	Usually cut with a sickle at 15–25 cm above the ground, PHT: Threshing, drying, winnowing and cleaning, grading, storage curing, milling, polishing	Cob sheath tums brownish, grains become hard. They are piled up for 24 h and then dried in the sun for 5 or 6 days to reduce the moisture to 10–12%	Most of the high. yielding varieties and hybrids mature in about 100–115 days, less than 25% moisture	Grain becomes hard containing 20–25% moisture
13.	Yield	*Transplant*: 70–80 q/ha grain, 80–85 t/ha straw *Aerobic*: 50–55 q/ha grain, 60–62 t/ha straw *SRI*: 80–85 q/ha grain, 90–95 t/ha straw	Grain 70–80 g/ha Straw 12 15 q/ha Queen of Cereal: High yield potential	*IR*: Grain 40–50 q/ha, Straw 10–12 t/ha *RF*: Grain 20–25 g/ha, Straw 57 t/ha	*IR*: Grain: 30–35 q/ha straw: 1 t/ha *RF*: Grain: 12–20 q/ha straw: 0.7 t/ha

Table 4.2 Production practices of Kharif fiber crops

S.No	Fiber	Cotton	Jute
1.	Botanical name Chromosome no.	*Gossypium hirsutum* $2n = 52$	*Corchorus* spp. $2n = 14$
2.	Soil type	Clay loam, black cotton soil, red soil, pH = 5.5–8	Loamy, sandy loam, pH 5–8 , old alluvial soil
3.	Climate	Germination at 15 °C Growth 21–25 °C Fruit and ball development 27 °C	Warm humid tropical subtropical crop, high humidity required; 1500 mm rainfall; 24–37 °C is optimum
4.	Land preparation	Deep ploughing, solarization, 1 ploughing followed by crisscross harrowing, planking	Deep ploughing followed by 2–3 crises cross harrowing.
5.	Sowing time	Irrigated: May 15 Rainfed: June- 15 to July 15	March to May
6.	Seed rate	Desi: 10–20 kg/ha American: 12–15 kg/ha	Broad: 10 kg/ha Drill: 6–8 kg/ha
7.	Seed treatment	100–120 days, 3–4 picking, done when boll fully burst and cotton hanged down	Treatment with the fungal culture of *Trichoderma viride* (local strain) exhibited a beneficial effect in jute fiber production
8.	Spacing	Desi: 60 × 15 cm 60 *\x 30 cm American: 60 × 15 cm, 90 × 60 cm	20 × 6 cm, 30 × 10 cm
9.	Fertilizer management	15–20 t/ha FYM irrigated: 100:50:50 Kg/ha NPK, rainfed: 50:50:25 kg/ha NPK application of PGR NAA helps in branching and bud formation	4–5 t/ha FYM 60:40:40 kg/ha NPK 15 kg urea foliar spray
10.	Water management	Sensitive to water deficient and water logging. First pre-sowing and second 3–4 week after germination	450–500 mm rainfall, first pre-sowing and 2–3 post sowing
11.	Weed management	Glyphosate and 2,4-D for control of existing weeds prior to planting	Hand weeding twice on 20–25 DAS and 35–40 DAS. Fluchloralin can be sprayed at 3 days after sowing at the rate of 1.5 kg per hectare and is followed by irrigation. Further one hand weeding can be taken up at 30–35 DAS.
12.	Harvesting	100–120 days, 3–4 picking, done when boll fully burst and cotton hanged down	110–120 days
13.	Yield	Irrigated 15–20 q/ha Rainfed: 10–12 q/ha	20–25 q/ha fiber

Table 4.3 Production practices of Kharif legumes

S.No	Legume	Groundnut	Soybean
1.	Botanical name Chromosome no.	*Arachis hypogaea* $2n = 20$	*Glycine max* $2n = 40$
2.	Soil type	Sandy loam, light soils easy to growth, pH 5.5–8 high acidic/alkaline harmful	Well drained fertile loam soils with pH 6.0–7.5. Water logging is injurious
3.	Climate	Tropical, long-warm weather during growth, avg rainfall 500–1200 mm, Best temperature: 21–27 °C; require warm and dry during ripening	Optimum temperature is 26.5–30 °C; grows well in warm and moist climate
4.	Land preparation	To ensure good surface tilth and a depth of 15 cm, plough the soil two or three times at the optimal soil moisture level. After each ploughing, use planking to keep moisture in the soil. Utilize a power tiller and a better plough	Prepare the land to get fine tilth and form beds and channels
5.	Sowing time	Kharif: May 15 to June 15 Rabi: August 15 to September 30	Kharif: June to August Rabi: September to November Summer: February to March
6.	Seed rate	Bunchy/early 100–115 kg, spreading 70–80 kg, drilling/dibbing method can be used	80 kg/ha
7.	Seed treatment	Treated with Captan or Thiram at 3 gm/kg. Chlorpyriphos is used @ 250 ml/45 kg of seed to prevent the seed damage from soil insects at initial stages	Carbendazim or thiram @ 2 g/kg of seed
8.	Spacing	30 × 10 cm spreading 45–60 × 15–20 cm	Dibble the seeds adopting spacing of 30 × 5 cm and for rainfed 30 × 10 cm. Depth of sowing is 2–3 cm
9.	Fertilizer management	FYM10-12 t/ha; NPK 20:50:25/ha $ZnSO_4$ 20–22 kg @ alternate year increases pod yield	Apply 20 kg N, 80 kg P205 and 40 kg K20 along with 40 kg of S. Give 40 kg P205 as foliar spray (2% DAP) on 40 DAS
10.	Water management	Very low requirement 450–500 mm sensitive to excess soil moisture, 3–5 irrigation @ different critical stages	Irrigate immediately after sowing. Irrigate at interval of 7–10 days during summer and 10–15 days in winter
11.	Weed management	Weeds controlled effectively with alachlor at the rate of 5 L in 500 L of water per hectare as a pre emergence soil spray within two days of sowing groundnut	Fluchloralin 2.0 L/ha or pendimethalin 3.3 L/ha 3 DAS. Fluchloralin 1 kg ai/ha or alachlor 2.0 kg ai/ha

(continued)

Table 4.3 (continued)

S.No	Legume	Groundnut	Soybean
12.	Harvesting	110–120 days, yellow vine a shedding, seed kernels turns brown	Yellowing of leaves and shedding indicates maturity
13.	Yield	*Bunchy:* 15–20 q/ha *Spreading:* 25–30 q/ha	18–20 q/ha

Table 4.4 Production practices of Kharif sugar crops

S.No	Sugar Crop	Sugarcane
1.	Botanical name Chromosome no.	*Saccharum officinarum L.* $2n = 60, 80$
2.	Soil type	Cultivated in a variety of soil types. loam that are moderately heavy and deep ($1–2m$) are preferable than deeper and heavier soils. The soil needs to be well-drained and deep
3.	Climate	A long warm growing season with adequate rainfall
4.	Land preparation	Tillage (ploughing, harrowing, subsoiling, etc.) Levelling also needs to be done. Incorporation of organic manures is proved to be beneficial
5.	Sowing time	Spring: February to March Autumn: September to October
6.	Seed rate	50,000 three budded setts, 75,000 two budded setts, 87,500 single budded setts
7.	Seed treatment	Before planting, seed sets are dipped for 10 min in a 0.5% Agallol (3%) solution or a 0.25% Aretan (6%) or Tafasan (6%) solution
8.	Spacing	Row spacing: 90–150 m and 240 cm
9.	Fertilizer management	Apply FYM/compost during field preparation Inorganic fertilizers: NPK 270: 112.5: 60 kg/ha. N and K applied in three equal quantities. N may be coated with neem cake @ 20%. Soils deficient in iron: 100 kg ferrous sulfate/ha and for zinc: 37.5 kg $ZnSO_4$
10.	Water management	Irrigate the crop as needed throughout the crop's life cycle Phase of germination (0–35 days) To improve germination, provide shallow soaking with 2–3 cm depth of water at shorter intervals, especially in sandy soil
11.	Weed management	Deep ploughing controls weeds On 3–4 days after planting, apply 1.75 kg of Atrazine. Sprayer for knapsacks on at 21 DAP, apply Gramoxone 2.5 L + 2–4, D sodium salt 2.5 L/ha as a directed spray. Weeding by hand before each manuring proves effective
12.	Harvesting	It takes 12–18 months to mature depending on the type and planting timing
13.	Yield	70–80 t/ha

Table 4.5 Production practices of Kharif pulses

S.No	Pulse crop	Moong/green gram
1.	Botanical name Chromosome no.	*Vigna radiata* $2n = 22$
2.	Soil type	Well-drained loam or sandy loam soils are ideal; alkali soils are not preferred. The pH range runs from 6.5 to 7.5.
3.	Climate	Optimum temperature is 28–30 °C. Annual rainfall: 60–75 cm.
4.	Land preparation	Prepare the land for fine tilth and the form beds and channels. Apply lime at a rate of 2 t per acre together with 12.5 t per acre of FYM or composted coir pith to increase yield by 15–20%
5.	Sowing time	*Kharif:* June to July *Rabi:* September to October
6.	Seed rate	Seed rate: 20 kg/ha for pure crop, 10 kg/ha for mixed crop, and for bund sowing 50 g/100 m length
7.	Seed treatment	Carbendazim or thiram @ 2 g/kg of seed, *Trichoderma viride* @ 4 g/kg or *Pseudomonas fluorescens* @ 10 g/kg
8.	Spacing	30 × 10 cm
9.	Fertilizer management	*FYM:* 12.5 t/ha fertilizer *Rainfed:* 12.5 kg N + 25 kg P205/ha *Irrigated:* 25.0 kg N + 50 kg P205/ha
10.	Water management	Irrigate every 10–15 days, depending on soil moisture. The flowering and pod development stages are critical stages
11.	Weed management	Spray Fluchloralin 1.5 L/ha or Pendimethalin 2.0 L/ha as pre-emergence 3 DAS followed by one hand weeding 30 DAS
12.	Harvesting	When 80% of the pods have matured, harvest the plants and stack them for a few days before the planting process
13.	Yield	*Rainfed:* 700–900 kg/ha *Irrigated:* 1500 kg/ha and

4.2.3 Production Practices of Zaid Crops

Table 4.10 shows the production practices Zaid crops: cowpea and black gram.

MCQs

1. Botanical name of Rice is:

   - *Pennisetum americanum*
   - *Gossypium hirsutum*
   - *Corchorus* spp.
   - *Oryza sativa*

2. Chromosome number of Jowar

   - 24

Table 4.6 Production practices of Rabi cereal crops

S.No	Cereal	Wheat	Barley
1.	Botanical name Chromosome no.	*Triticum aestivum* $2n = 20$	*Hordeum vulgare* $2n = 14$
2.	Soil type	Soils with a clay loam or loam texture, good structure, and moderate water holding capacity are ideal for wheat cultivation	Barley is salt and alkalinity tolerant but acidity sensitive
3.	Climate	Optimal temperature is 20–22 °C	Optimum temperature is 20–22 °C
4.	Land preparation	Waterlogged regions should be avoided at all costs. The land should be ploughed extensively three to four times to ensure that the soil is well pulverized and properly levelled	One ploughing and two or three harrowing by tractor or bullock force. In places where termites are a concern, mix Benzene Hexa Chloride 10% at 20–25 kg/ha or aldrin 5% dust at 10–15 kg/ha into the soil
5.	Sowing time	October 15 to first week of November	Mid-October to Mid-November
6.	Seed rate	75–100 kg/ha *Broadcast:* 125 kg/ha	Irrigated: 100 kg/ha Rainfed: 80100 kg/ha
7.	Seed treatment	Carbendazim or Thiram at 2 g/kg	Carbendazim or Thiram at 2 g/kg
8.	Spacing	20–22.5 cm apart	Spacing 22.5 cm row for irrigated with 5 cm depth 22.5–25 for rainfed with 6–8 cm depth
9.	Fertilizer management	Spread 12.5 t/ha of FYM 4–6 week before sowing *IR:* 120:60:40 kg/ha *RF:* 40:30:20 kg/ha Basal dose: 50% N + full P&K 30 DAP 50%N	*FYM:* 12.5 t/ha *Irrigated:* 60:30:20 kg NPK/ha *Rainfed:* 40:20:20 kg NPK/ha (100% NPK as basal)
10.	Water management	There are five crucial watering stages: immediately after sowing (15–20 DAS), crown root initiation, active tillering (35–40 DAS), flowering (50–55 DAS), and grain fullness (70–75 DAS)	Water is required in the range of 200–300 mm. 2–3 irrigation. Seedling, active tillering, flag leaf, and milking or soft dough stages are critical stages
11.	Weed management	For broad leaf weeds, apply herbicides 2,4-D at 0.75-1 kg a.e/ha or Isoproturon at 800 g/ha as a pre-emergence spray and three days after sowing-two hand weedings are essential	Pendimethalin (pre-emergence) 1.0 kg/ha or post-emergence herbicides Isoproturan @ 0.75 kg/ha + 0.5 kg 2,4D at 3–5 leaf stage + one hand weeding were found to be efficient weed control methods

(continued)

Table 4.6 (continued)

S.No	Cereal	Wheat	Barley
12.	Harvesting	The leaves and stems turn yellow and become quite dry. Moisture content is 20–25%. Harvest with a sickle. The grain is dried for 3–4 days to achieve a moisture content of 10–12%	Animal or mechanical threshers are used for threshing followed by winnowing and cleaning. Grain storage at 10–12% moisture level is the optimum
13.	Yield	*IR:* 60–70 q/ha *RF:* 12–15 q/ha	Grain: 3.0–3.5 t/ha Straw : 4.0–5.0 t/ha

- 36
- 20
- 12

3. Yield of jute per hectare is ______kg/ha?

- 40–50
- 20–25
- 35–40
- 60

4. Botanical name of groundnut is:

- *Gossypium hirsutum*
- *Corchorus spp.*
- *Arachis hypogaea*
- *Vigna radiata*

5. Spacing in Moong is:

- 20 × 25 cm
- 25 × 35 cm
- 30 × 10 cm
- 20 × 40 cm

6. Irrigated yield of wheat is?

- 60–70 q/ha
- 70–100 q/ha
- 20–30 q/ha
- 40–50 q/ha

7. Sowing time of lentils is ______

- First fortnight of June
- Second fortnight of October
- First fortnight of December
- None of the above

Table 4.7 Production practices of Rabi pulses

S.No	Pulses	Pea	Lentils
1.	Botanical name Chromosome no.	*Pisum sativum* $2n = 14$	*Lens culinaris* ssp. $2n = 14$
2.	Soil type	Soil having a pH range of 6.0–7.5. Highly susceptible to water logging	Crop can withstand moderate alkalinity. Acidic soils are unsuitable
3.	Climate	Pea is a cool-season crop; the ideal temperature for seed germination is 22 °C	Needs a chilly environment and is sown in the winter. The ideal temperature for growth is between 18 and 30 °C
4.	Land preparation	Prepare a well-pulverized seed bed	Plough land with a plough. Cross harrowing and planking need to be done. Raised bed planting is more effective than flat bed
5.	Sowing time	Mid-October to mid-November	Second fortnight of October
6.	Seed rate	*Field peas:* 60–80 kg/ha *Garden peas:* 100–125 kg/ha	30–40 kg/ha
7.	Seed treatment	To avoid seed-borne illnesses, treat the seeds with 4 g/kg Trichoderma or 2 g/kg Thiram or Captan. 2 kg of Rhizobium culture is applied to the seeds, and 2 kg Phosphobacterium is applied to the soil soon before sowing	To avoid seed-borne illnesses, treat the seeds with 4 g/kg Trichoderma or 2 g/kg Thiram or Captan. 2 kg Rhizobium culture is applied to the seeds, and 2 kg Phosphobacterium is applied to the soil soon before sowing
8.	Spacing	Garden pea: 20 cm Field pea: 30 cm	30 cm row to row
9.	Fertilizer management	FYM should be applied at a rate of 20 t/ha with 60 kg N, 80 kg P, and 70 kg K (/ha) as a base and 60 kg N/ha 30 days after sowing	N: 20–25 kg/ha P: 50–60 kg/ha
10.	Water management	Once in a week	1–2 irrigation *First irrigation:* 40 DAS *Second irrigation:* at flowering (or) pod formation
11.	Weed management	Fluchloralin 0.75 kg ai/ha or hand weeding twice	Fluchloralin 0.75 kg ai/ha as pre-planting spray (or) hand weeding twice at 30 DAS and 60 DAS
12.	Harvesting	Harvest can be done on 75 days after sowing. High temperature during harvest affects the quality of peas	When the plants dry up and moisture reaches 10–12%
13.	Yield	8–12 t/ha of pods can be obtained	1.8–2 t/ha

Table 4.8 Production practices of Rabi forage crop

S.No	Sugar crop	Oats
1.	Botanical name Chromosome no.	*Avena sativa* $2n = 42$
2.	Soil type	Except for waterlogged soils, oats can be grown in every soil type
3.	Climate	Optimum is cool and wet
4.	Land preparation	Soil should be ploughed to fine tilth by multiple ploughings
5.	Sowing time	Mid-September to Mid-December
6.	Seed rate	100 kg/ha
7.	Seed treatment	Treatment should be done with (Carboxin 37.5% + Thiram 37.5% DS) 2 g/kg seed to ensure freedom from covered smut disease
8.	Spacing	Lines should be 20 cm apart
9.	Fertilizer management	Apply FYM 12.5 t/ha 80:40:0 kg NPK/ha. Basal dressing of 40 kg N and 40 kg P2O5 should be done at the time of seeding and rest 40 kg N should be applied as top dressing each after first and second cutting
10.	Water management	It only takes three to four irrigations. If there are many cuttings, the field needs to be irrigated after each one
11.	Weed management	Only one hand weeding needs to be done
12.	Harvesting	Crop needs about 120–150 days to mature
13.	Yield	Fodder: 50–60 t/ha and grain: 400 kg/ha

8. *Helianthus annus* is the scientific name of?

- Mustard
- Linseed
- Jute
- *Sunflower*

9. Seed rate of *Pennisetum americanum is*

- 3–5 kg/ha
- 20–25 kg/ha
- 10–15 kg/ha
- 25–35 kg/ha

10. Sunflower is harvested in _______ days

- 100
- 80
- 120
- 60

Table 4.9 Production practices of Rabi oilseeds

S.No	Oilseed	Sunflower	Mustard	Linseed
1.	Botanical name Chromosome no.	*Helianthus annus* $2n = 34$	*Brassica nigra* $2n = 36$	*Linum usitatissimum* $2n = 30$
2.	Soil Type	Grows best in soils with pH neutral to slightly alkaline soils range from 6.5 to 8.0. Complete failure is seen in 4.6 pH sandy soil	Soil types range from sandy loam to clay, thrives well in light soil, Mustard on any soil except waterlogged ones and alkaline soils. pH 6.5–7.5, neutral soil is ideal	Central India's deep cotton soil, North India's alluvial loam soil, and soil must be well drained
3.	Climate	During the germination and seedling stages, the crop needs a chilly climate; from seedling up until flowering, it needs warm weather. Flowering time shouldn't fall during a period of heavy rain or drizzle	In India, it is a crop grown during the rabi season. The ideal temperature is between 18 and 25 °C, with cool, dry weather. High RF, high humidity, and cloudy weather are unfavorable during flowering. Its most vulnerable to frost	Needs cooler climate for proper growth and development. Temperature should be 25–30 °C during germination, 15–20 °C during seed formation, but fiber requires still lower temperature
4.	Land Preparation	Until all the clods are broken and a fine tilth is obtained, plough once with a tractor, twice with an iron-plough, or three to four times with a country-plough	Given that seeds are small, a fine seed bed is necessary. The use of a flatbed is also necessary for fertilizer and seed drilling	Fine and smooth seed bed free from clods. The soil should be fee of termites and ants
5.	Sowing Time	Irrigated sunflower can be sown from: December to January or April to May	Rabi: October and continues till December end	For rainfed crop, the optimum time for sowing is Mid-September to mid-October
6.	Seed Rate	For hybrids in case of rainfed the seed rate should be 5 kg/ha and irrigated 4 kg/ha	Seed rate varies from 4 to 7 kg/ha	20–30 kg/ha for line sowing, 35–40 kg/ha for broad casting
7.	Seed Treatment	For rainfed sowing, it is advised to soak seeds in 2% ZnSO4 for 12 h before drying them in the shade. Use 2 g/kg of carbendazim or thiram on the seeds Prior to planting, treat the seeds for 24 h	treat seed with Apron 35 SD @ 6 g/kg seed. Mancozeb @ 3–4 g/kg seed or Thiram/Captan/Carbendazim 2 g/kg	Bavistin 2 g/kg seed

(continued)

Table 4.9 (continued)

S.No	Oilseed	Sunflower	Mustard	Linseed
8.	Spacing	Hybrids: 60 cm × 30 cm Varieties: 45 cm × 30 cm	45 × 30 cm in beds	20–30 cm row to row
9.	Fertilizer Management	Before the final ploughing, uniformly distribute 12.5 t/ha of FYM, compost, or composted coir pith throughout the field. Fertilizer dose of NPK should be 82 kg/ha N, 13 kg/ha P, 60 kg/ha K, 9.4 kg/ha S, 37 kg/ha Ca and 21 kg/ha Mg	Basal: FYM 25 t N 25 kg, P 60 kg. Top dressing: N 25 kg Sulfur is removed in large and needs returned to soil	Irrigated: 30–40 kg of N and P Rainfed: 20 to 30 kg/ha of N and P Relay cropping: 10–15 kg/ha of N (Relay cropping of linseed is done in UP, India)
10.	Water Management	Irrigate immediately after sowing, then again on the fourth or fifth day, and then at intervals of seven to eight days, depending on the soil and weather conditions, as well as the seeding, blooming, and seed development stages, that is, two weeks before and after flowering	The total amount of water needed is 400 mm, and moisture during the pre-flowering and pod-filling stages is crucial. Two irrigations are required. Three irrigations are needed for light sandy soils	Two irrigations at 35 and 75 DAS. In light soils, 3–4 irrigations
11.	Weed Management	Apply Pendimethalin as pre-emergence spray three days after sowing, or apply Fluchloralin at 2.0 l/ha as pre-emergence spray five days after sowing, followed by watering. On the 15th and 30th days after sowing, hoe and hand weed to get rid of the weeds	Pendimethalin pre- emergence 0.5–1.5 kg/ha based on soil type. Fluchloralin 1.25 kg pre-plant incorporation and post emergence Isoproturan 0.75 kg/ha is used	Up to 25 DAS, there should be weed-free condition in the field 2 hand weedings are done at 21 DAS and 35–40 DAS
12.	Harvesting	Harvest in 120 days	Duration: Three to four months. After the pods turn brown, the plants are removed, dried in the sun, and then threshed. Maturity in 30–40 days	Crop should be harvested at the red ripe stage for the fiber crop and for grain crop, at physiological maturity Storage moisture: 10–12% Oil content in seed: 36–42%

(continued)

Table 4.9 (continued)

S.No	Oilseed	Sunflower	Mustard	Linseed
13.	Yield	*Rainfed:* 6–8 q/ha *Irrigated:* 15–17 q/ha	Irrigated rapeseed 1.5–2.0 t/ha Rainfed rapeseed 1.0–1.5 t/ha Irrigated mustard 2.0–2.5 t/ha Rainfed mustard 1.5–2.0 t/ha	Irrigated grain yield: 1–1.2 t/ha

Table 4.10 Production practices Zaid crops

S.No	Oilseed	Cowpea	Black gram
1.	Botanical name Chromosome no.	*Vigna unguiculata* $2n = 22$	*Vigna mungo L.* $2n = 22$
2.	Soil type	Well drained loam or sandy loam preferred. Alkali soils are not suitable. Optimum pH range is 6.0–7.5	Black gram can be grown in a wide range of soil types, from sandy to heavy cotton soils. A well-drained loam with a pH of 6.5–7.8 is suitable. Alkaline and saline soils are not suitable
3.	Climate	Prefers temperatures between 27 and 35 °C for optimum growth. Can withstand drought to some extent	Temperatures between 25 and 35 °C are ideal for growing, but it can withstand temperatures as high as 42 °C Black gram is a tropical crop that thrives in hot, humid climates Heavy rains during the flowering period are damaging to the crop because they limit crop production
4.	Land preparation	Three or four ploughings should be used to finely tilth the soil, and field should made in 1 feet wide ridges and furrow	Prepare the land to fine tilth and form beds and channels for sowing
5.	Sowing time	October to November	Mid-June is the optimum time to sow black gram
6.	Seed rate	20–25 kg/ha	20–30 kg/ha
7.	Seed treatment	Before sowing, treat the seeds with carbendazim or thiram (2 g/kg) or *Trichoderma viride* (4 g/kg) or pseudomonas fluorescens (10 g/kg)	Seed treatment with Carbendazim 1 g + Thiram 1.5 g per kg of seed
8.	Spacing	30 × 10 cm	Row-to-row distance should be 30 cm

(continued)

Table 4.10 (continued)

S.No	Oilseed	Cowpea	Black gram
9.	Fertilizer management	12.5 t/ha FYM Rainfed : 12.5 kg N + 25 kg P2O5 + 12.5 kg K2O +10 kg S/ha Irrigated: 25 kg N + 50 kg P2O5 + 25 kg K2O + 20 kg S/ha (App of sulfur is subject to use of single super phosphate for phosphorus)	20:40:20 NPK kg/ha along with 20 kg S/ha (App of sulfur is subject to use of single super phosphate for phosphorus)
10.	Water management	After sowing, irrigate immediately. Irrigation is a must throughout the critical flowering and pod-formation stages. Avoid water logging	At critical growth stages, flowering and pod formation, irrigation should be provided
11.	Weed management	One hand weeding and hoeing between 20 to DAS. Pre emergence application of Pendimethalin has proven effective	Weeding and hoeing 25–30 DAS and second weeding at 45 DAS Pendimethalin or Metalachlor @ 1.0–1.5 kg/ha are found very effective as post emergence
12.	Harvesting	The crop duration is of 75–90 days. Plants are uprooted, dried by stacking and then threshed to remove grains	Maturity indices include 70–80% pods maturity and most of the pods turn black. Rain and over maturity leads to shattering
13.	Yield	Yield range from 0.5 to 0.8 t/ha. Green fodder yield is 10 t/ha	12–15 q/ha of grains is obtained

References

Krushna Bhokare. – Package of Practices for Field Crops
Tamil Nadu Agriculture University. – AgriTech Portal

Chapter 5
Weed Management

Abstract Weeds are one of the major problems faced by modern-day farmers. They compete with the crops for various resources and also harbor pests and diseases. Thus, weeds pose a great threat to the crop production. This chapter explores various facets of weed management. Various topics like classification of weeds, crop-weed interaction, effects of weeds, and most importantly the management of weeds for sustainable crop production have been discussed.

Keywords Weeds · Competition · Herbicides · Allelopathy · Fumigants · Eradication · Herbigation · Fumigation

5.1 Introduction to Weeds

Weeds are plants that are unwelcome in a particular environment and can be harmful, dangerous, or economically damaging. It is a plant that is out of place and is not intentionally sown. More yield loss and production expenses are incurred by farmers due to weeds than by insect pests, crop diseases, and nematodes worldwide. Weeds considerably contribute to the deterioration of land and water, decrease farm and forest output, and displace native species.

Weeds are the plants that grow where they are not wanted (Jethro Tull, 1731).

- In 1967, the Weed Science Society of America defined a weed as "a plant growing where it is not desired" (Buchholtz, 1967).
- In 1989, the Society's definition was changed to define a weed as "any plant that is objectionable or interferes with the activities or welfare of man welfare of man" (Humburg, 1989, p. 267; Vencill, 2002, p. 462).
- The European Weed Research Society defined a weed as "any plant or vegetation, excluding fungi, interfering with the objectives or requirements of people" (EWRS, 1986).

L. Ahmad et al., *Fundamentals and Applications of Crop and Climate Science*,
https://doi.org/10.1007/978-3-031-61459-0_5

- The Oxford English Dictionary (Little et al., 1973) defines a weed as a "herbaceous plant not valued for use or beauty, growing wild and rank, and regarded as cumbering the ground or hindering the growth of superior vegetation."

Weed invasion continues to pose a serious threat to both the integrity of our natural ecosystems and the productive potential of land and water, despite the millions of dollars invested by various governments and private agencies across the world on the issue. The management of weeds is a crucial part of plant protection, increasing the potential for crops to produce more. It includes weed management in such a way that the crop maintains its production potential while not being harmed by the weeds. Weed control is accomplished through mechanical, cultural, and chemical means. Biological control methods in field crops are being considered, but are not widely used, although the use of herbicides in weed control is fairly common throughout the world.

5.2   Characteristics of Weeds

Around 3%, or 8000 species, of the roughly 250,000 plant species that exist in the globe act as weeds. Only 200–250 of the 8000 are of significant importance with respect to global farming systems. If a plant has particular traits that distinguish it from other plant species in the surroundings, then it is regarded as a weed (Cahoon et al., 2018).

The following features of weeds enable them to persist and proliferate in nature:

1. Abundant seed production.

   While most crop plants only generate a few hundred seeds per plant, weeds can produce tens of thousands or even millions of seeds each plant. For example, Pigweed 117,000.

   seeds produced, *Amaranthus viridis*, 11 Million seeds produced.

2. Seed dormancy.

   Dormancy is essentially a phase of rest or a brief condition during which the weed seed does not germinate for a variety of reasons like temperature, moisture, pH, oxygen level, light, etc. For example, summer annual weed seeds typically do not germinate in the autumn, protecting them from the harsh winter weather.

3. Seed dispersal.

   Weed seeds have a remarkable ability to spread from one location to another by wind, water, and other creatures, including people. Due to their size, weed seeds frequently get confused with crop seeds and travel from one area to another with them, leading to their easy dispersal.

4. Seed viability.

   Weed seeds remain viable for longer period without losing their viability, for example, annual meadow grass (*Poa annua*), creeping thistle (*Cirsium arvense*) for 20 years, and field bind weed (*Convolvulus arvensis*) for about 50 years.

5. Capacity to withstand adverse/unfavorable conditions by the virtue of enhanced germination rate, faster multiplication capacity, etc.
6. Weeds are more exhaustive than most of the field crops. They extract almost all essential nutrients that our crop needs for survival and development.
7. Resistance to pests and diseases Weeds as we know are wild in nature hence show the property of high biotic and abiotic resistance leading to their better survival during the times of distress and unfavorablity.

5.3 Harmful and Beneficial Effects of Weeds

Harmful Effects
- Weeds compete with the crops for nutrients, space, light, oxygen, water, etc. If deprived of these essential components, our crop fails to give us yields, thereby leading to monetary losses.
- The effects of weeds on agricultural output are significant. In most developed nations, weeds are thought to reduce agricultural output by 5%, in the third world nations by 10%, and in least developed nations by 25%.
- Among the total annual losses of agricultural product from various pest, weeds account for 45%, insects for 30%, diseases 20%, and other pest 5% as shown in Fig. 5.1 (Rao, 2000).
- In addition, weeds serve as an alternative host for insects, pests, diseases, and other microbes.

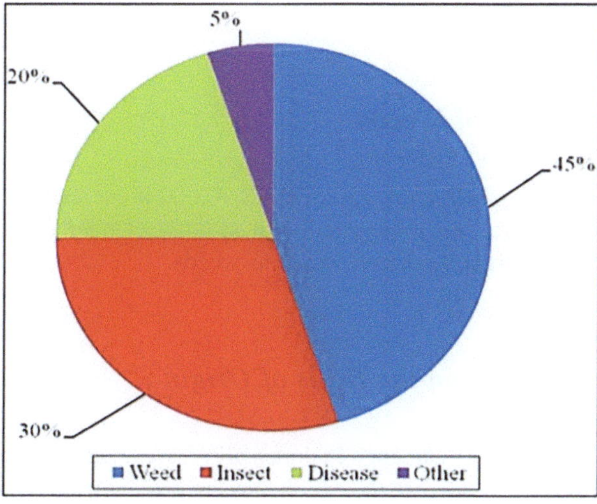

Fig. 5.1 Agricultural losses. (Rao, 2000)

- High intensity of weed leads to slow germination and initial growth, wider row spacing, and slow lateral spread, which causes tremendous loss in productivity as well as quality (Alka Sahrawat et al., 2020).
- In addition to lowering productivity, weeds disrupt agricultural operations. Weeds make mechanical sowing, harvesting, and other mechanized operations challenging, resulting in greater costs for labor, equipment, and chemicals to remove them.
- Some weeds release poisonous substance inhibitors into the soil, which can be harmful to crop plants, humans, and livestock. Weeds cause human health problems like hay fever, asthma, itching, inflammation, etc.

Beneficial Effects

- Weeds contribute to soil moisture retention and erosion prevention. Weeds will provide a ground cover that will lessen the amount of exposed bare soil, which will aid to conserve nutrients, notably nitrogen, that may otherwise be leached away, especially on light soils.
- Leaves of several species of *Amaranthus* eaten as a relish throughout East Africa, often mixed with salt and groundnut paste. Seed can be roasted and pounded to produce flour used in preparation of bread and biscuits (Alka Sahrawat et al., 2020).
- For birds, weeds can be a valuable source of food. It might play a key role in increasing the bird populations.
- Weeds have been proved for producing nectar for the bees and other pollinating species.
- Weeds like dandelion are a delicacy for some human ethnic groups in South Asia. Some weeds are of medicinal importance too.
- Weeds can also be valuable indicators of growing conditions in a field, for example, they can be excellent indicators for pH levels, organic matter, nutrients, moisture content, etc.

5.4 Classification of Weeds

Weeds account for approximately 250 of the 3,90,000 plant species and are prevalent in both agricultural and nonagricultural systems. Under current conditions, approximately 30,000 species are classified as weeds.

5.4.1 Classification on the Basis of Origin

- *Indigenous Weeds*: This category includes all native weeds of the country. The majority of weeds are native to the nation they are present in. Weeds like *Acalypha indica and Abutilon indicum* are native to India.

- *Exotic Weeds*: These weeds are non-native to a country and are mostly invasive in nature. They are introduced from other countries. They are very problematic, and their control is very tedious and problematic. For example, weeds like *Lantana, Eichhornia, and Parthenium* were all brought to India. In the jungles of Uttar Pradesh and Madhya Pradesh, *Lantana camara* has displaced a number of species. Water hyacinth, or *Eichhornia crassipes*, has choked wetlands and other bodies of water, killing a number of aquatic plants and animals.

5.4.2 Classification on the Basis of Morphology

- *Grasses*: All weeds belonging to the Poaceae family are referred to as grasses because of their distinctively tall, thin, spiny leaves. Examples include *Cynodon dactylon* and *Echinochloa colonum.*
- *Sledges*: This group includes the weeds from the Cyperaceae family. The majority of the leaves grow from the base of the modified stem, either with or without tubers. Examples include *Fimbristylis miliacea* and *Cyperus rotundus.*
- *Broad-Leaved Weeds*: The majority of weeds fall under this category, with the exception of those that were already covered. Broad-leaved weeds are all dicotyledonous weeds. Examples include *Tridax procumbens, Digera arvensis*, and *Flaveria australasica* (Fig. 5.2).

5.4.3 Classification on the Basis of Season

- *Annual Weeds*: Annual weeds are weeds that only exist for a season or a year and complete their life cycle during that season or year. These are small herbs with weak stems and shallow roots. They produce a lot of seeds, and seeds are usually the mode of propagation. After sowing, annuals wither away after the season ends, but the seeds grow and produce the following season or year's crop. Annual

Fig. 5.2 Cynodon dactylon, Cyperus rotundus, and Digera arvensis

Fig. 5.3 *Chenopodium album* and *Commelina benghalensis*

Fig. 5.4 Sonchus arvensis, Sorghum halepense, and Convolvulus arvensis

weeds are the most prevalent in fields, for example, *Chenopodium album* and
Commelina benghalensis (Fig. 5.3).

- *Biennial Weeds*: In the first season, it completes its vegetative growth. The following season, it flowers, produces seeds, and then eventually dies. These are
  primarily found in undisturbed natural settings, for example, *Daucus carota.*
- *Perennial Weeds*: Perennials have a life cycle of at least 2 years and sometimes
  nearly indefinitely. They change to withstand stressful conditions. They spread
  through underground stem, root, rhizomes, tubers, and other means in addition to
  sexually via seeds, for example, *Sonchus arvensis, Sorghum halepense,* and
  Convolvulus arvensis (Fig. 5.4).

5.5   Interaction of Weeds and the Crop

Competition is a mutually negative interaction between two individual plants that
require the same limited resources, such as nutrients, water, and light. When two
such individuals grow in close spatial proximity, competition for limited resources

leads to a reduction in survival, growth, or reproduction of both individuals (Harper, 1977). The inherent characters and adaptations that a weed possesses show that the interaction of crop and weed will almost at all times be negative, given a few exceptions where surprisingly the interaction comes out to be beneficial for one or for both. The characters that weed possesses that advocates for this negative includes the ability of weed to extract more nutrients from the soil, producing large numbers of seed with favorable dormancy, which provides them viability for prolonged periods of time, rapid establishment and growth of seedlings, tolerance to shading effects by the crop or by other weeds and the ability to reproduce more frequently.

Crop and weeds compete with one another naturally in order to grow and produce as much as possible, with the development of each species sometimes coming at the price of the other. It happens when the amount of moisture, nutrients, light, and possibly carbon dioxide needed by the plants exceeds the available supply. Competition between crop and weed, as well as between specific plants of each, may arise. The eventual result of competition typically leads to the formation of a distinctive crop-weed association.

As a result of the weed's suppression, the crop's yield is decreased. Row crop cultures also very frequently experience the weed suppression caused by the crop. The early plants to settle a region have an advantage over the later ones, according to a theory of plant competition. In actual weed control, where cropping procedures are always intended to establish the crop before the weeds, this idea is of utmost importance.

5.5.1 Crop-Weed Competition

It is rightly said by a famous scientist, "one year seeding, seven years weeding," which means that if a weed invades an area, it is going to take the farmer 7 years of continuous eradication and management procedures to keep the weed under the manageable threshold. If kept unsupervised, weeds can very easily wipe off the crop because of adaptations like better stress resistance, better seed dispersal, frequent and easy propagation, and nutrient extracting capabilities.

In the crop weed competition, the weed competes with crop in the following ways:

- *Competition for nutrients.*

- Weeds typically absorb mineral nutrients more quickly than many crops and store a disproportionately higher amount of them in their tissues. For example, *Amaranthus* sp. accumulate over 3% N on dry weight basis and are termed as "nitrophills." *Chenopodium* sp. and *Portulaca* sp. are "K" lovers with over 1.3% K_2O in dry matter *(Weed Management-TNAU)*. In each crop season, nitrogen loss from weeds is significant and frequently twice as much as that from crop plants. For instance, weeds were found to extract nine times more N, ten times more P, and seven times more K during the early phases of maize cultivation.

- *Competition for light.*
- If a dense weed growth smothers the crop seedlings, competition for light may start very early in the crop season. When there is an abundance of moisture and nutrients, it becomes a significant component of crop-weed competition. In years with moderate rainfall, the crop-weed competition in dry land agriculture is confined to nitrogen and light. When weeds shade a crop plant, it faces competition for nutrients and moisture but not enhanced light intensity.
- According to Berti and Sattin (1996), the main competitive force for weeds growing taller than crop may be the shading produced by the leaves of weeds located above the canopy. This finding suggests that the competitive effect of weeds appears to be primarily tied to other relative cover.
- *Competition for moisture.*
- In general, weeds absorb more water than the majority of our field crops do for producing an equivalent quantity of dry matter. As soil moisture stress increases, as is the case in arid and semiarid regions, this competition becomes more and more important.
- From water use efficiency point of view, most of weed species are C4 type having higher WUE as compare to the C3 type of crop species (Silva et al., 2007). Generally speaking, C4 plants use water more effectively, producing more biomass per unit of water. The transpiration rate of *Cynodon dactylon* was almost two times that of pearl millet.
- *Competition for space (CO_2).*

- The need for CO_2 necessitates crop-weed competition for space, and this competition may arise in conditions of densely packed plant communities. A more effective way to use CO_2 by C4 type weeds may speed up their growth relative to our C3 type crops (Table 5.1).

5.5.2 Allelopathy

Allelopathy is a biochemical phenomenon that occurs when one plant species' chemical has inhibitory and/or stimulatory effects on another plant species (Molisch, 1937; Rice, 1984).

Table 5.1 Critical period for crop weed competition in different field crops

S.No.	Crops	Days from sowing	S.No.		Days from sowing
1.	Rice (Lowland)	35	7.	Cotton	35
2.	Rice (upland)	60	8.	Sugarcane	90
3.	Sorghum	30	9.	Groundnut	45
4.	Finger millet	15	10.	Soybean	45
5.	Pearl millet	35	11.	Onion	60
6.	Maize	30	12.	Tomato	30

The word "allelopathy," which covers both direct and indirect effects of exuded chemicals on a target plant, is broad (Weston, 1996). Allelochemicals, or exuded chemicals, are secondary metabolites that, upon release, have an impact on the development, survival, and reproduction of the target plants. Allelochemicals, which directly affect the target plant regardless of the surrounding abiotic or biotic environment, mediate direct allelopathic effects (Inderjit & Weiner, 2001). Biotic agents like microbes and lower plants that metabolically convert plant components into chemical forms that are frequently harmful to the target plant transmit indirect allelopathic effects (Dakshini et al., 1999).

Allelochemicals, which are phenolic acids, flavonoides, and other aromatic compounds including terpenoids, steroids, alkaloids, and organic cyanides, are synthesized by plants as end products, by-products, and metabolites that are released from the plants.

- *Allelopathic effects of weeds on crop plant.*

- In case of Maize *(Zea mays)*, leaves and inflorescence of *Parthenium* sp. affect the germination and seedling growth.

- In case of Maize, tubers of *Cyprus rotundus* affect the dry matter production of wheat.
- *Allelopathic effects of crop plant on weeds.*
- Chemical exudations from the root of Maize inhibit the germination of *Chenopodium album.*
- *Allelopathic effect of weeds on other weeds.*

- *Digitaria sanguinalis* and *Amaranthus* spp. are restricted in their growth by an inhibitor released by decomposing rhizomes of *Sorghum halepense.*

5.6   Principles of Weed Control

Weed control is one of the most important practices that needs to be done in case of agricultural crops. If weeds are allowed to grow unsupervised, they can easily eradicate our main crop because of its inherent features. Before selecting a proper method of weed control, one must have a deep understanding about the weed that needs to be controlled, like the seed dispersal method, life span, propagation method, frequency of propagation, dormancy of seed, etc. Along with the knowledge of weeds, one must also have the proper understanding of the crop and the soil that has been infested by weeds. Keeping all these things in consideration, only then a proper step can be taken on controlling the weeds.

Principles of weeds control include prevention, eradication, control, and management.

5.6.1   Prevention

It includes all the methods undertaken to prevent the introduction or establishment of weeds in a given area. If sufficient preventive steps are not implemented to lessen weed infestation, no weed management program will be successful. Long-term planning is included in order to manage or control weeds more efficiently and economically than is possible when they are allowed to spread unchecked. Some preventive methods include checking for physical purity of crop weeds and removal of weed seeds if at all present, keeping irrigation channels, canals and sources free from weed seeds, regular inspection of farm for weeds, quarantine measures for avoiding the entry of weed seeds in a country and inspecting the farm implements for weed seeds, and other impurities before using.

5.6.2   Eradication

It implies that a certain weed species, along with its seeds and vegetative parts, have been eradicated or entirely removed from a specific area and that the weed won't grow again unless new plants are introduced to the area. Eradication is typically only tried in smaller areas, such as a few hectares or a few thousand square meters or fewer, due to its difficulties and high expense. Eradication is frequently utilized in high-value areas including greenhouses, decorative plant beds, and containers. When the weed species is so toxic and persistent that it makes cropping difficult and expensive, this may be desirable and economical.

5.6.3   Control

It includes procedures that minimize but do not always completely eradicate weed infestations. When using management methods, the crop produces a normal yield, and the weeds' growth is severely constrained but seldom eradicated. In general, the characteristics of the included weeds and the efficacy of the employed management method determine the degree of weed control attained.

5.6.4   Management

Weed management is a system approach whereby whole land use planning is done in advance to minimize the very invasion of weeds in aggressive forms and give crop plants a strongly competitive advantage over the weeds. Weed control aims at only getting rid of the weeds present by some kind of physical or chemical means, while weed management is a system approach.

Weed management is divided into four categories, which when used in a combined manner yields the best results: cultural methods, physical methods, biological methods, and chemical methods.

5.6.4.1 Cultural Methods

Cultural weed control comprises nonchemical crop management strategies such as variety selection, land preparation, harvesting, and postharvest processing. If carried out appropriately, these methods greatly assist in weed control. While cultural approaches cannot completely eradicate weeds, they can help to reduce their population. As a result, they ought to be employed in addition to other strategies. To create favorable conditions for the crop, a variety of cultural activities including solarization, mulching, tillage, planting, fertilizer application, and irrigation are used. In addition, factors including variety choice, sowing window, cropping strategy, and farm hygiene can all help reduce weed growth. Techniques for crop and cultural management provide crops the healthiest chance of fending off weeds.

Advantages of Cultural Methods
- They are easy to practice.
- They don't require much skills, and they can be done by layman too.
- They are cheap and affordable to most of the farmers.
- They are very effective.
- They don't pose any threat or damage to the main crop.
- They are easy to adopt.
- They don't leave out any residue problem like other methods.

Disadvantages of Cultural Methods
- They are slow and time-consuming in nature.
- They do not eradicate the weeds completely.
- Weeds that are perennial are difficult to control by these methods.

5.6.4.2 Mechanical/Physical Methods

Mechanical or physical methods either eliminate weeds or create an environment that is less conducive to weed survival and seed germination. Since man first started cultivating crops, mechanical or physical weed control methods have been used. Tillage, mowing, hoeing, hand weeding, sickling, burning, floods, mulching, and other mechanical processes are among them.

A small number of weeds can be effectively controlled by digging or pulling. In tiny garden spaces or flower beds, smaller weeds are readily hand-pulled or hoed out; however, larger woody weeds may require the use of heavy machinery.

Mulching utilizes a physical barrier to limit light and prevent weed development, making it a mechanical control method as well.

Advantages of Mechanical Methods
- These methods are age old tried and tested.
- They do not harm or damage our main crop in any means.
- Removal of weeds from in-between the crops becomes easy.
- With the help of machinery, larger area can be covered in shorter periods of time.

Disadvantages of Mechanical Methods
- The cost of machinery is high sometimes.
- Machines, if used inappropriately, can damage the crop.
- These methods are usually labor-consuming along with time-consuming.
- Use of machinery can be expensive.

5.6.4.3 Biological Methods

It involves using other living things, such as insects, pathogens, snails, or even competitor plants, to suppress weeds. Weeds cannot be completely eliminated with the biological control method, although their population can be decreased. Not all weed varieties can be controlled with this strategy. The best targets for biological control are introduced weeds. A program for integrated weed management works well with biological control, in theory. However, biological control has some drawbacks, including the fact that it is a lengthy process, its results are not always instant or sufficient, only few weeds are suitable candidates, and earlier biological control attempts have experienced a relatively high failure rate.

Advantages of Biological Methods
- Biological methods of weed control are long-lasting.
- They do not harm the environment.
- They do not leave any residue.
- They are relatively cheap if looked at from a longer perspective of time.

Disadvantages of Biological Methods
- These methods take a long time to show effects.
- They don't eradicate the weeds completely.
- They are costly.
- The multiplication rate is low.
- There are fairly lesser number of success stories as compared to other methods.

5.6.4.3.1 Bio-Herbicides

Bio-herbicide use is the talk of the town in recent times. They include the application of plant pathogens that are designed to eradicate the desired weed.

These are artificially cultivated native pathogens that are sprayed on target weeds every season in agriculture regions, much like post-emergence herbicides. Since bacteria and viruses can't actively enter the host and need natural openings or vectors to spread disease to plants, fungi are more commonly utilized as weed pathogens than bacteria, viruses, or nematodes.

5.6.4.4 Chemical Methods

Chemical methods include the use of herbicides extensively. Herbicides are compounds that are used to control weeds. Herbicides are chemicals that safeguard crops (or desirable plants) by killing weeds or obstructing natural plant growth. Herbicides offer a practical, affordable, and efficient way to aid with weed management. They enable for earlier planting dates, less ploughing during planting, and more time for other operations needed for the farm. Reduced tillage has resulted in a decrease in soil erosion, which has decreased from approximately 3.5 billion tons in 1938 to one billion tons in 1997. As a result, less soil is entering streams, which improves the quality of the country's surface water. Agriculture that uses no till is impossible without the usage of herbicides.

However, using herbicides also entails dangers for the environment, the ecosystem, and human health. Before choosing the best control, it is crucial to comprehend both the advantages and downsides of chemical weed control.

On some farms, herbicides may not be required, but mechanical and cultural weed techniques become even more crucial in these situations. Herbicides come in a wide variety of varieties. The best times, places, and methods to employ a specific herbicide depend on a variety of circumstances.

Advantages of Chemical Methods
- They act fast.
- They are easy to use.
- They kill the unwanted plant in the matter of days.
- They are cheaper when compared to mechanical counterparts.
- Only one-time application is enough, no recurrent applications are needed.

Disadvantages of Chemical Methods
- They have detrimental effect on environment.
- They pollute the groundwater.
- Chemicals can enter the food chain and can be lethal for various animals including human beings.
- They can also affect our main crop if not used and sprayed properly.
- They are expensive if the covered area is huge.
- In some cases, exposure to herbicides has shown to reduce the fertility rates in women.

More on herbicides will be discussed in the coming sections. First, we will take a look at integrated weed management.

5.6.4.5 Integrated Weed Management

Integrated weed management combines various agronomic practices to manage weeds, reducing reliance on any single weed control technique. The combination of two or more weed-control methods at low input levels to reduce weed competition in a given cropping system below the economical threshold level is known as integrated weed management.

The integrated weed management approach aims to reduce residue problems in plants, soil, air, and water. An IWM entails the use of a combination of mechanical, chemical, and cultural weed management practices in a predetermined sequence so as not to disrupt the ecosystem. Based on national research, corn and soybean yield can be reduced by approximately 50% without effective weed control (Singh, 2020).

IWM is essential in modern agriculture. It is so because one method of weed control may be effective and economical in one situation but not in another. Furthermore, using only chemical controls over a long period of time causes significant problems for the environment and humans, particularly the issue of weed resistance to a specific herbicide, and no single herbicide is effective in controlling the diverse flora that weeds exhibit. If IWM is not practiced, the use of a single approach may result in an increase in the population of a specific weed.

The *FAO* gave the definition of IWM as, "It is a method whereby all economically, ecologically and toxicologically justifiable methods are employed to keep the harmful organisms below the threshold level of economic damage, keeping in the foreground the conscious employment of natural limiting factors."

A good IWM system should have high practicability for farmers, good economic viability, and scientific feasibility. It should extend beyond the limits of the farm and identify the root causes of weed spread, which possibly could be roadsides, no cultivated land, irrigation canals, etc.

Advantages of IWM
- IWM prevents weed shift toward perennial nature.
- Suitable for high cropping intensity.
- It does not cause environmental pollution.
- It poses no danger of herbicide residue in soil or plant.
- It gives higher net return.
- It prevents development of herbicide resistance in weeds.

5.7 Herbicides

The selectivity of certain chemicals to cultivated crops in controlling associated weeds without affecting the crops serves as the foundation for chemical weed control. Such selectivity may be due to morphological differences, differential absorption, differential translocation, or differential deactivation.

5.7.1 Classification of Herbicides

5.7.1.1 Based on Method of Application

- *Soil Applied Herbicides*: They are directly incorporated in the soil. They act through underground parts of the weed, for example, Fluchloralin.
- *Foliar Applied Herbicides*: They act on the foliage of the weed, for example, Glyphosate and Paraquat.

5.7.1.2 Based on Time of Application

- *Pre-Plant Application*: These herbicides are applied before the sowing of our main crop. Both the soil and foliar application are taken into consideration. For example, Glyphosate can be applied on the foliage of perennial weeds like *Cyperus rotundus* before planting of any crop.

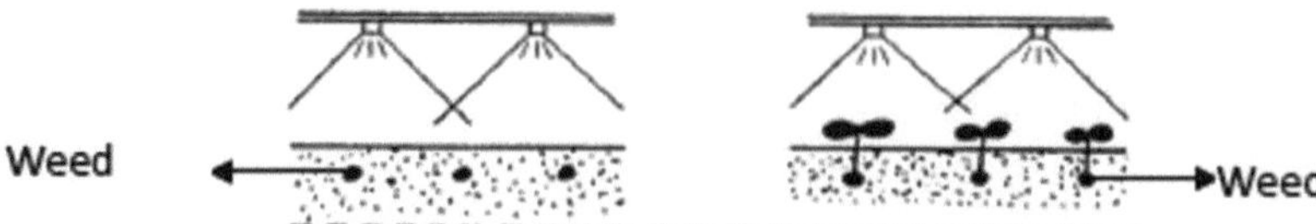

- *Preemergence*: Herbicide is applied before the emergence of crop or weed itself. In the case of annual crops, application is done after crop sowing but before weed emergence, and this is referred to as crop preemergence, whereas in the case of perennial crops, it is referred to as weed preemergence. For example, Atrazine soil application by spraying to sugarcane on the third day after sowing can be referred to as pre-emergence to sugarcane crop, whereas pre-emergence to weed is soil application by spraying Atrazine immediately after a rains to check any new growth of weeds in an inter-cultivated field (Butachlor, Pendimethalin, Atrazine).

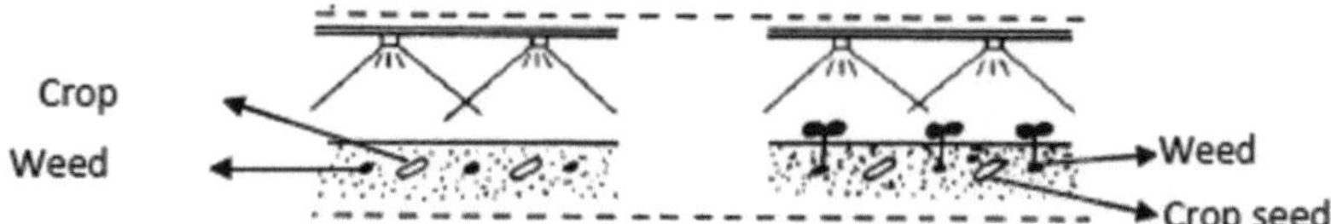

- *Postemergence*: Postemergence herbicide application is done after the emergence of a crop or weed. Early postemergence occurs when weeds appear before crop plants emerge from the soil and are controlled with herbicide. For example, postemergence spraying of 2,4-D Na salt is used to control the parasitic weed striga in sugarcane. (Paraquat, Glyphosate, 2,4-D Na salt).

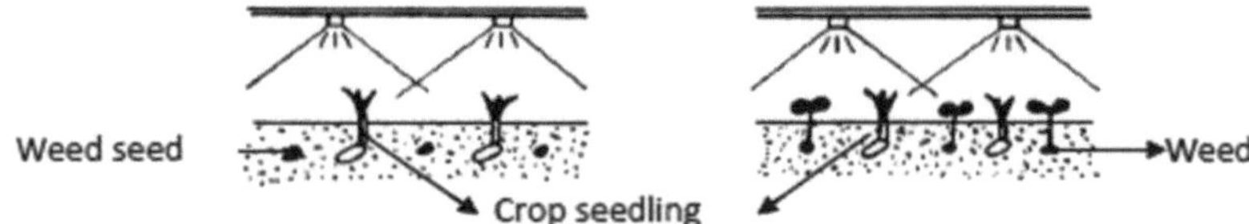

- *Early postemergence*: Early postemergence refers to the application of Paraquat to control emerging weeds after 10–15 days following sowing of potato.

5.7.1.3   Based on Mobility

- *Contact Herbicides*: It kills the plant when it comes in the direct contact of the plant, for example, Paraquat.
- *Translocated Herbicides*: They travel up to the plant via the vascular tissues and kill it from the inside, for example, Glyphosate.

5.7.1.4   Based on Mode of Action

- *Selective Herbicide*: This herbicide when sprayed, to a mixture of plant species, will affect only a single one and will leave others unharmed, for example, Atrazine.
- *Nonselective herbicide*: It destroys all the vegetation it comes in contact of, for example, Paraquat.

5.7.2   *Method of Application of Herbicides*

There are two major methods of applying herbicides to the field. They are further divided into several types, briefly shown in Table 5.2.

1. *Soil Application.*

   - *Surface Application*: Spraying or broadcasting is used to evenly apply soil-active herbicides to the soil's surface. The herbicides that have been sprayed are not disturbed or incorporated into the soil. Herbicides are incorporated to avoid volatilization and photodecomposition. For example, Fluchloralin is incorporated in rainfed conditions and left undisturbed under irrigated conditions.
   - *Subsurface Application*: Herbicides are applied in a concentrated band around 7–10 cm below the soil's surface in order to control perennial weeds, for example, Carbamate herbicides to control *Cyperus rotundus*.

Table 5.2 Methods of herbicide application (e-Krishi Shiksha)

	Soil application		Foliar application
1.	Surface	i.	Blanket spray
2.	Subsurface	Ii.	Directed spray
3.	Band	Iii.	Protected spray
4.	Fumigation	Iv.	Spot treatment
5.	Herbigation		

- *Band Application*: Application to a restricted band along the crop rows, leaving a wider region between the rows untreated. Later, the weeds are eradicated by cultivating the inter rows. Here, cost savings are feasible.
- *Fumigation*: Fumigation is the process of applying volatile chemicals to soil or enclosed areas to create a gas that can kill weed seeds. Fumigants are the name for herbicides used during fumigation. These work effectively to get rid of weed seeds as well as perennial weeds, for example, methyl bromide.
- *Herbigation*: It involves applying herbicides together with irrigation water using sprinkler and surface irrigation systems. Farmers in India use fluchloralin most of the times.

2. *Foliar Application.*

- *Blanket Spray*: Herbicides are uniformly applied to standing crops without taking the location of the crop into account. Here, only very specific herbicides are used, such as spraying rice with 2,4-ethyl ester three weeks after transplanting.
- *Directed Spray*: It involves spraying herbicides on weeds that are growing between rows of crops while avoiding the crops themselves. By using a shield or hood for protection, this would be achieved. For instance, spraying Glyphosate with a hood between rows of Tapioca to manage *Cyperus rotundus*.
- *Protected Spray*: This technique involves covering the widely spread crops with polyethylene covers or similar materials before spraying nonselective herbicides to control weeds. This is both costly and time-consuming. However, farmers are employing this kind of glyphosate spraying to manage weeds.
- *Spot Treatment*: To eradicate and stop the spread of weeds, it is typically applied to limited areas with bad weed infestations.

MCQs

1. How many seeds are produced by *Amaranthus viridis?*

- 11 Million
- 9 Million
- 17 Million
- 21 Million.

2. What percentage of plant species in the world act as weeds?

    - 4%
    - 3%
    - 8%
    - 11%.

3. The total annual losses of agricultural product from various weeds account for what percent?

    - 30%
    - 70%
    - 45%
    - 12%.

4. Which one of the following species of weeds is native to India?

    - *Echinochloa colonum.*
    - *Lantana camara.*
    - *Acalypha indica.*
    - *Eichhornia crassipes.*

5. All the grasses belong to which family?

    - Poaceae.
    - Fabaceae.
    - Verbenaceae.
    - Amaranthaceae.

6. Which one of the following is a broad leaf weed?

    - *Fimbristylis miliacea.*
    - *Echinochloa colonum.*
    - *Tridax procumbens.*
    - *Cyperus rotundus.*

7. *Chenopodium album* is an example of?

    - Annual weed.
    - Perennial weed.
    - Biennial weed.
    - None of the above.

8. Chemical exudations from the root of Maize inhibit the germination of _____.

    - *Chenopodium album.*
    - *Cyperus rotundus.*
    - *Echinochloa colonum.*
    - *Acalypha indica.*

9. Sickling is which type of weed control?

- Chemical.
- Mechanical.
- Cultural.
- Biological.

10. Example of an important post emergence herbicide is.

- Glyphosate.
- Atrazine.
- Butachlor.
- Pendimethalin.

References

Berti, A., & Sattin, M. (1996). Effect of weed position on yield loss in soybean and comparison between relative weed cover and other regression models. *Weed Research, 36*, 249–258.

Buchholtz, E. (1967). *Agricultural practices and innovations*. Publisher.

Cahoon, C. W., et al. (2018). *Weed control in field crops* (Vol. 5, pp. 1–20).

Dakshini, K. M. M., Foy, C. L., & Inderjit. (1999). Allelopathy: One component in a multifaceted approach to ecology. In K. M. M. Dakshini, C. L. Foy, & Inderjit (Eds.), *Principles and practices in plant ecology: Allelochemical interactions* (pp. 3–14).

European Weed Research Society. (1986). *Proceedings of the 6th EWRS Symposium*. Publisher.

Govt. of Manitoba. *Integrated weed management: Making it work on your farm*.

Harper, J. L. (1977). *Population biology of plants*. Academic Press.

Humburg, K. (1989). *Advances in crop science*. Publisher.

Inderjit, & Weiner, J. (2001). Plant allelochemical interference or soil chemical ecology? *Perspective in Plant Ecology, 4*, 3–12.

Little, T. M., Hills, F. J., & Carmer, S. G. (1973). *Statistical methods in agricultural research*. Publisher.

Molisch, H. (1937). *The influence of one plant on another: Allelopathy*.

Rao, V. S. (2000). *Principles of weed science* (pp. 427–436).

Rice E. L. (1984). *Allelopathy*. 422 pp.

Sahrawat, A., et al. (2020). The potential benefits of weeds: A comparative study: A review 149–150.

Silva, A. A., Ferreira, F. A., Ferreira, L. R., Santos, J. B. (2007). Biologia de Plantas daninhas. In *Topicos em Manejo de Plantas Daninhas* (pp. 17–61).

Singh, A. (2020). Integrated weed management. *Weed Research, 2*.

Tull, J. (1731). *The horse-hoeing husbandry*. Publisher.

Vencill, W. K. (2002). *Herbicide handbook*. Weed Science Society of America.

Weston, L. A. (1996). Utilization of allelopathy for weed management in agro ecosystems. *Agronomy Journal, 88*, 860–866.

Chapter 6
Water Management

Abstract Excessive, indiscriminate, and unscrupulous use of water has resulted in the crisis of water everywhere. Usable water is a critical output for agriculture. Therefore, there is an immediate need to devise a thoughtful water management plan in order to utilize the potential of every droplet of water toward successful crop production. This chapter dwells deep into core concepts of water management like water resources of India (with special reference and focus on India), plant water relations, functions of water, irrigation and fertigation, droughts and floods, and water harvesting.

Keywords Critical stage · Lysimeter · Fertigation · Rainfed · Watershed · Drainage · Droughts · Floods

6.1　Importance of Water in Agriculture

It is a well-known and widely accepted fact that agriculture is in direct dependence on water. Water is of critical importance to agriculture. Water quality is therefore essential for agriculture, both for the safety and quality of the output as well as for the sustainability of the agriculture industry.

Although water covers a sizeable chunk of the earth's surface, there is a very small part of it (3%), which is suitable for human use. Out of this, only 1% is fully exploitable for human purposes, majorly agriculture (70% of all the water available for human use is utilized by agriculture), because the rest is trapped in polar glaciers (UN, 1997).

The irrigated agriculture accounts for 20% of the total cultivated land (World Bank, 2022), and it contributes 40% to the food produced in the world. Since the 1960s, with the advent of Green Revolution, the use of high-yielding varieties, intense irrigation, and better plant protection measures has led to better nutritional and food security throughout the world. But at the same time, environmental degradation in the form of excessive water use, groundwater depletion, and human health hazards have been on an exponential graph since then.

In agriculture, the water is majorly used for irrigation purposes, drinking water for livestock, washing down livestock building, etc. However, because it does not immediately replenish water into the ecosystem, this method of water consumption is demanding.

Some of the water used for irrigation on farms relies on rivers and streams, either via surface extraction or borehole extraction. Some crops demand low water quantity throughout their life cycle like strawberries, but most field crops require large quantity of freshwater for proper growth and development.

Good crop yields and the production of livestock are the results of efficient and safe water use in agriculture. However, poor water quality can have a severe impact on the growth of crops, and ultimately economic performance.

Irrigation water contamination is also one major problems that is encountered on daily basis by the farming community. These contaminants in water can frequently be the cause of persistent fungal and bacterial infections in plants. Water used for agricultural needs to be cleaned up before usage. This will support the maintenance of plant and crop quality. Any disinfectant can be used, given the fact that it should be effective against a broad spectrum of germs as conceivable. Pathogens can be found on irrigation equipment too. For instance, water recirculation systems can disperse mold spores throughout the entire greenhouse crop; hence, they also need regular checkups and sanitation done on them.

6.2 Water Resources in India

India gets all of its water through precipitation. India's annual precipitation, including snowfall, is estimated to be 1200 mm, which equates to 4000 km^3 or 400 million hectare-meters (mha-m) (Kumar et al., 2005). Rivers, canals, reservoirs, tanks, ponds, lakes, and brackish water are examples of surface water bodies in India.

India has 1.2 million ha of irrigation canals and 1.9 million ha of reservoirs. Nearly 2.2 mha is the area under the village ponds and tanks.

Precipitation is the main source of replenishable groundwater. Only 50 of the 400 mha of rainwater percolates into the ground and joins the other 215 mha of groundwater that are usable (Fig. 6.1).

6.3 Function of Water in Plants

Water plays several crucial physiological roles in plants. Water is essential to plant life. It is necessary for plants in the below mentioned ways:

- Water contains two essential elements, oxygen and hydrogen, which are required for carbohydrate synthesis during photosynthesis.

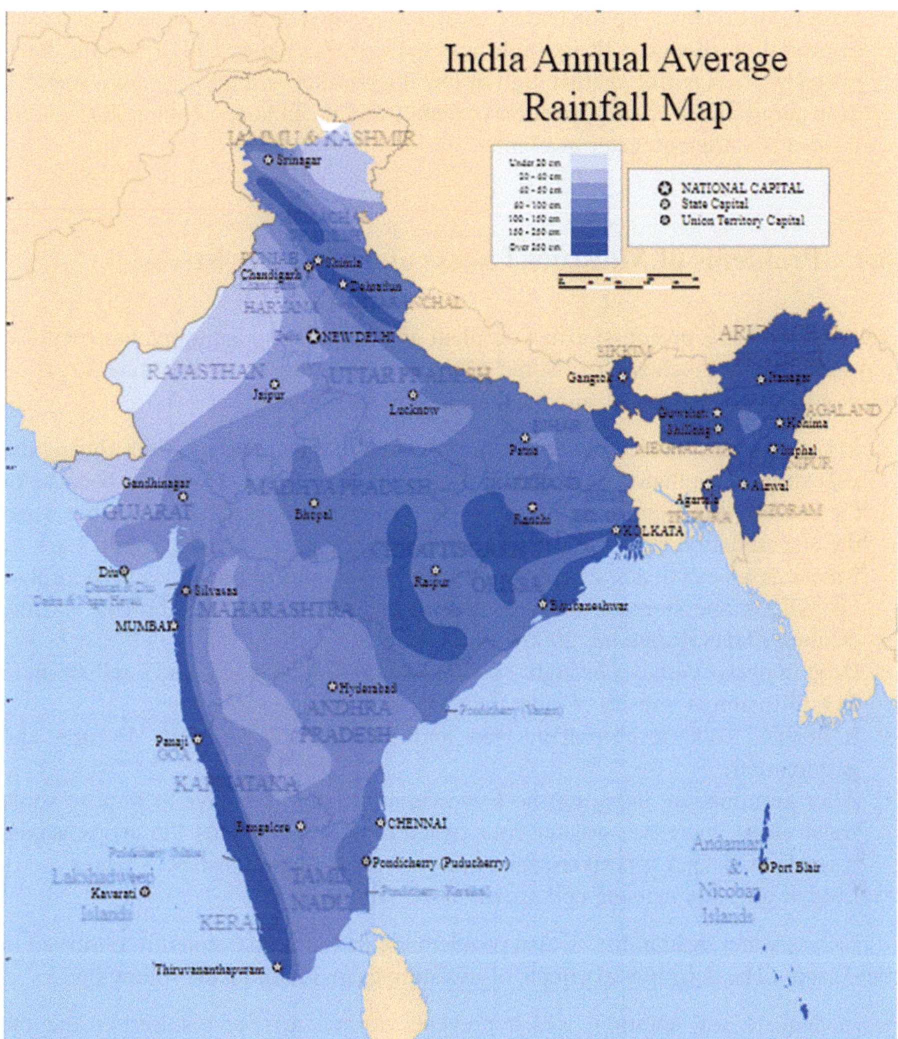

Fig. 6.1   Annual Rainfall in India

- Water is a structural component of plant cells that maintains cell shape via turgor pressure. When there is plenty of water, cells become turgid, and plants retain their structural form. Water makes up the majority of an actively growing plant's body weight, accounting for 85–90% of the body weight of young plants and 20–50% of the body weight of older or mature plants.
- Water acts as a solvent and a medium in plants, allowing metabolic reactions to take place. It also acts as a solvent for plant nutrients and aids in nutrient uptake from soils.

- Solar radiation heats up the leaves. Heat is dissipated by plants through increased transpiration. Because of its high heat by vaporization and high specific heat, water acts as a buffer against high or low temperature injury.
- Transpiration is a crucial function in plants, and it can happen at a potential rate if there is sufficient water available.

6.4 Problems of Moisture Excess and Moisture Stress

Water stress has a profound effect on plant growth, altering anatomy, morphology, physiology, and biochemistry. Some of the negative impacts of moisture stress on plant growth, development, and harvest include the following:

- Moisture stress has less of an impact on yield and its yield qualities during the vegetative stage than it does during the reproductive stage. However, because of the stress at this point in the plant's growth and development, photosynthesis and the accumulation have the greatest impact on the plant (Kabiri, 2010). Lack of water causes cell death, disruption of plant physiological processes such as photosynthesis and stomatal closure, and alterations in plant metabolism and the death of plants (Rahmani, 2006).
- Drop in photosynthesis brought on by a reduction in leaf area and a reduction in CO_2 diffusion as a result of stomata being closed to preserve water.
- It impacts root development, fruit and leaf growth, and cell division and germination.
- After germination, delaying the first irrigation for a few days to impose some water stress increases greater root penetration, which helps the crops access water from deeper soil layers and cope with drought circumstances better.
- Loss of turgidity causing cell growth to be inhibited.

Excess water availability is also problematic and is very frequently observed in field crops. The detrimental effects of moisture excess include the following:

- Inadequate soil aeration, which prevents normal aerobic respiration and the activity of microorganisms.
- Wet conditions coupled with warm temperatures creates congenial platform for fungal growth and seed germination is affected (Rajanna et al., 2019).
- Root penetration is restricted by a high water table, which also hinders crop development and productivity.
- Additionally, it leads to the buildup of soluble salts in the root zone.

6.5 Plant-Water Relations

All living cells contain a significant amount of water, which serves as a medium in which all substances are dissolved and go through a variety of reactions. Water itself is both a reactant and a product. It serves as a mode of transportation. Water is necessary for cellular growth and turgidity. Watermelon has 92% water, while the majority of herbaceous plants have 85–90%. In trees and shrubs, the woody parts contain a comparatively smaller amount of water compared to the softer parts, which contain a significant amount. A seed may appear dry, but it contains 10–15% water. It will stop breathing and die without this water.

A plant depends heavily on water for its survival. A total of 500 g of water are absorbed by the roots, transported to the plant body, and lost to the atmosphere for every gram of organic matter produced by the plant. Water deficiency and severe cellular process malfunction can result from even minor imbalances in this flow of water.

6.5.1 Plant-Water Potential

The three main elements that govern the energy status of water in plant cells are turgor pressure (p), imbibitional pressure (m), and solute or osmotic pressure (s).

The phrase "turgor pressure" can comprise pressures brought on by both intercellular pressure and gravitational forces. The total water potential in plants can be stated as follows:

$$\Psi = \psi_p + \psi_m + \psi_s$$

Where

ψ = total water potential,
Ψp = turgor potential.
Ψm = imbibitional potential.
Ψs = solute or osmotic potential.

6.5.1.1 Relative Water Content

The ratio of actual water content to water content at saturation (fully turgid) is known as relative water content (RWC), and it is typically expressed as a percentage. The dry weight (DW) of the sample is subtracted from the fresh weight to determine its actual water content (FW).

The difference between saturation weight, also known as turgid weight (TW), and dry weight is the amount of water present at saturation.

In irrigation water management, the following are the major areas of water-plant relationships: water absorption and water loss or transpiration.

6.5.1.2   Absorption of Water by Plants

Water absorption refers to the process by which plants, more specifically plant root hairs, absorb water. It is a crucial process for plants because they need to absorb both water and minerals to survive. Higher plants have specific portions of their roots that absorb water, whereas lower plants' entire bodies function as a water-absorbing absorb.

- Capillary water from the soil is primarily absorbed by plants.
- Root hair is the primary organ for absorbing water. Root hairs are protuberances of the epidermal cells with thin walls.
- To get to the leaves, the absorbed water travels through a variety of cells, including cortical cells, endodermis, pericycle cells, and xylem tubes.
- Major roles for absorbed water in transpiration, photosynthesis, cell vitality, turgidity, and rigidity.

6.5.1.2.1   Process of Water Absorption in Plants

Water absorption in plants can be classified into two types based on the involvement of metabolic energy.

1. Absorption by active means.
2. Absorption by passive means.

Active absorption is the absorption of water by using metabolic energy, usually against a concentration gradient (energy from ATP).

Passive absorption: Water absorption that does not require the use of metabolic energy.

6.5.1.2.2   Factors Affecting Water Absorption

Several factors influence water absorption, including the following:

(a) The amount of water in the soil.
(b) Temperature of the soil.
(c) Soil solution concentration.
(d) The amount of air in the soil.
(e) Transpiration.
(f) Root systems.
(g) Metabolism.

6.5.1.2.3   Conduction of Water in Plants

Water and minerals are absorbed by plant roots. The root hairs that are present there maximize the surface area of the roots for absorbing water and the minerals dissolved in it. The water and dissolved minerals are taken up by the xylem in the roots from the root hair. Water moves up the stem through the xylem due to the transpiration pull, a force produced by transpiration in the leaves. Water and dissolved minerals (sap) are drawn through the stem by transpiration pull, which causes the sap to ascend. The root cells and xylem's conductive system are the routes used by liquid water as it moves from the soil to the leaf cells. It moves from leaf cells to the air as a vapor via intercellular gaps.

6.5.1.3   Transpiration

Mineral nutrients and water are absorbed by plants from the soil, but most of the water is not used by the plants. The process of transpiration allows the plants to release the extra water by evaporating it through various plant parts like stems or the stomata on the surface of the leaves. Water in plants can be drawn to great heights by a suction pull caused by the evaporation of water from leaves. In hot weather, transpiration aids in cooling the plant.

Typically, only about 5% of the water absorbed by plants is used for metabolic processes and body weight production, leaving the remaining 95% to be transpired. And 90–95% of the water lost from leaves is due to stomatal transpiration. Cuticular transpiration accounts for the final 5–10% (Vince Ördög 2011).

Furthermore, 90–95% of transpiration takes place during the day, and only 5–10% does so at night. The exception is the pineapple, a plant with a Crassulacean acid metabolism (CAM), in which the majority of transpiration occurs at night.

6.5.1.3.1   Factors Affecting Rate of Transpiration

Cellular factors (water status of the plant, leaf orientation, distribution and number of stomata in a leaf), environmental factors (light, relative humidity, temperature, wind velocity, atmospheric pressure), water availability, leaf surface area, and so on all influence the rate of transpiration.

6.6   Water Requirements of Agricultural Crops

Crop water requirements (CWR) are the volumes of water needed over a certain time period to make up for evapotranspiration losses from a cultivated field. The typical unit of measurement for crop water needs is millimeters per day, month, or season (Todorovic, 2005).

One of the fundamental requirements for crop planning on a field and for the design of any irrigation project is the assessment of the water requirement (WR) of crops. Water is primarily required to meet evapotranspiration (ET) and a plant's metabolic needs, which are collectively referred to as consumptive usage (CU). Since very little water is consumed by the plant during its metabolic processes, ET is essentially equated to CU.

Water requirements include the amount of water needed for particular operations like land preparation, as well as the losses resulting from ET (or CU), the application of irrigation water, and other losses. Considering these factors, we can form an equation, mentioned below (Table 6.1):

$$WR = ET \text{ or } CU + \text{application losses} + \text{special needs}.$$

6.6.1 Classification of Evapotranspiration

There are three types of evapotranspiration: potential, actual, and reference evapotranspiration.

Potential Evapotranspiration (PET): The word refers to the highest rate of evapotranspiration (ET) produced by a short, actively growing crop or vegetation with ample foliage that completely shades the ground surface and a plentiful supply of soil water in a particular environment.

Reference Evapotranspiration (ET_O): The word refers to the rate of evapotranspiration from a wide surface of 8–15 cm tall green grass cover of uniform height that is actively growing, fully shading the ground, and not water-stressed.

Actual Evapotranspiration (ET crop): It is used to describe the rate of evapotranspiration experienced by a certain crop during a specific time period under the current soil, water, and atmospheric conditions. When computing it from reference crop ET (ET_o) derived by various empirical equations or evaporation rates from evaporimeters, it makes use of a crop factor called crop coefficient. The ET crop varies according on the soil, water, and climatic conditions, as well as the stage of crop growth, location, and time of year.

Table 6.1 Water requirements of various field crops along with their water use efficiency and grain yield (Reddy & Reddy, 2004)

Crop	Water requirement (mm)	Grain yield (kg/ha)	Water use efficiency (kg/ha-mm)
Rice	1200	4500	3.7
Sorghum	500	4500	9.0
Pearl millet	500	4500	8.0
Maize	625	5000	8.0
Groundnut	506	4680	9.2
Wheat	280	3534	12.6
Finger millet	310	4137	13.4

6.6.2   Classification of Consumptive Use of Water

Peak Consumptive Use: Peak period consumption is the average daily consumption during a few days (often 6–10 days) of the highest consumption in a season.

Daily Consumptive Use: The amount of water consumed in a 24-h period is referred to as the daily consumptive use.

6.7   Critical Stages of Crop Growth for Irrigation

Water is essential to plant life and must be provided in the right quantities. The majority of the soils are irrigated in addition to receiving water from rain. The amount of soil moisture loss determines how long it takes between irrigations. Normally, the crop shouldn't be permitted to use more than 50% of the water that is available. In comparison to heavy soil, the intervals are shorter in sandy soil. The crops are only irrigated at critical periods when the water supply is extremely low.

The critical stage of water requirement is the point at which severe yield decline is caused by water stress. Additionally, it is called the moisture-sensitive period. The yield will be permanently decreased by moisture stress caused by a lack of water during the moisture sensitive period or critical stage. Even if enough water and fertilizer are provided during other growth stages, the output loss caused by stress at critical points won't be recovered.

Here are some important critical stages of crop growth for irrigation of various field crops:

- *Rice*: Submergence under water at tillering and flowering stages.
- *Wheat*: For tall wheat, the ideal soil moisture range is from the field's capacity to 50% of its potential. The ideal moisture range for dwarf wheat is between 100 and 60% of available moisture. Depending on the kind of soil, the crop's active root zone ranges from 50 to 75 cm. Crown-root initiation (3 weeks after sowing), flowering, and grain formation are the critical growth stages.
- *Maize*: On various soil types, the ideal soil moisture range is between 100 and 60% available in the maximum root zone, which measures 40–60 cm. The early vegetative period (30–40 days after sowing) and tasselling are the critical growth stages (45–50 days).
- *Pulses and Legumes*: Early vegetative growth, flowering, and pod development are critical phases for legumes.
- *Barley*: Early tillering, the boot stage, and grain filling are the critical stages for water requirements.

6.8   Introduction to Irrigation and Irrigation Scheduling

Irrigation is application of water to the crops artificially. It fosters crop growth. Rainfed agriculture is what was undertaken in the olden times, but with the advent of irrigation and irrigation scheduling, it has on a whole altered the landscape of how farmers grow the crops. Traditional rainfed agriculture is a high-risk operation even in regions with average seasonal precipitation that may seem enough since rainfall is typically irregularly distributed or soils have low water retention capacity.

Stable food production is made possible by irrigation. Irrigation can extend the growing season in some places. Irrigation has helped in improving food and nutritional security throughout the globe. Irrigation security makes it possible to afford additional inputs like better pest management, more fertilizer, higher yielding cultivars, and enhanced tillage (Table 6.2).

6.8.1   Irrigation Scheduling

In order to properly manage irrigation, water must be applied when the crop actually needs it, with only enough water to moisten the soil in the root zone. The main goal is to maximize agricultural productivity while using water in the most effective and economical way possible. This is the first and foremost idea of irrigation scheduling.

The availability of irrigation water and the crops' requirement for water are often the two main factors that determine when to irrigate. But the main factor in choosing when to irrigate crops is their need for water.

Since irrigation water is typically in short supply in most places, it must be used carefully and wisely. Thus, the idea of water economy is useful. Farmers are constrained to miss some irrigation in places with limited water supplies, which results in reduced yield due to the crops' poor growth and development. This generates financial losses for farmers alongside mental distress.

To prevent both over- and under-irrigation and to maintain high water use efficiency, irrigation timing should be based on crop needs.

Since crop irrigation demands are determined by the evaporative demand of the atmosphere, soil water status, and plant characteristics, it is crucial to have a

Table 6.2 Growth in irrigated land and population since 1900

Year	Area (million acres)	Population (billions)
1900	100	1.5
1950	235	2.5
1970	422	3.7
1990	598	5.3
1997	669	5.9
2017	806	7.5

Adapted from (FAOSTAT, 1999; FAO, 1998, 2021)

complete understanding of the link between soil, water, and atmosphere. The criteria for scheduling irrigation can be divided into three categories:

1. Criteria based on plants: It inculcates conditions of plants like, plant appearance, plant water potential and water content, plant growth, the critical crop stages of water need, stomatal aperture and leaf diffusion resistance, and plant temperature.
2. Criteria based on soil water status: It includes soil water content, critical level of available soil water, soil water tension, etc.
3. Criteria based on meteorological factors: It includes factors like frequency and interval of irrigation, irrigation period, natural form of precipitation a crop received, etc.

6.9 Types of Irrigation

There are four major types of irrigations, which can be used under various situations. No one type can be used in all the crops, with no consideration of topography and other factors. These four are as follows:

- Surface irrigation.
- Subsurface irrigation.
- Drip irrigation.
- Sprinkler irrigation.

6.9.1 Surface Irrigation

The most used irrigation technique is surface irrigation. In this technique, channels of varying sizes and shapes carry water directly onto the soil surface. On agricultural or orchard crops, as well as on moderately sloping terrain, surface irrigation systems are used. With surface irrigation, water runs over the soil's surface and is dispersed across the field. Surface irrigation techniques typically require less pressure than sprinkler irrigation, which results in cheaper operating costs per applied unit of water.

Its benefits include: management is simple; no modern technology is required. It can be accomplished with traditional knowledge. There is no requirement for capital and no energy costs. It is possible to do this on sloping land and in vast fields.

Surface irrigations are of various types such as the following:

- Border irrigation.
- Check basin irrigation.
- Wild flooding.
- Furrow irrigation.
- Basin irrigation (Figs. 6.2, 6.3, and 6.4).

Fig. 6.2   Furrow irrigation in Australia 2006

Fig. 6.3   Basin irrigation

6.9.2   Subsurface Irrigation

Subsurface irrigation, often known as subirrigation, is the practice of watering crops from below the soil's surface, either by digging trenches or setting up tile or perforated pipe lines. During the entire irrigation period, water is pumped into trenches and left to stand there so that it can travel laterally and vertically through the soil between the trenches via capillary action.

The primary requirements for subirrigation are (1) the presence of a high water table or an impervious subsoil above which an artificial water table can be formed and (2) highly permeable root zone soil with generally homogeneous texture allowing good lateral and upward transport of water. (3) Irrigation water is expensive and in short supply, and (4) soil salinity should not be an issue.

Fig. 6.4 Border irrigation

6.9.3 Drip Irrigation

Drip irrigation is one of the most water efficient form of irrigation. It drips water into the soil at very low rates (2–20 liters/hour) using a system of small diameter plastic pipes equipped with outlets known as emitters or drippers. Drip irrigation is sometimes referred to as trickle irrigation. In contrast to surface and sprinkler irrigation, which includes watering the entire soil profile, water is supplied near to the plants so that just the area of soil where the roots develop is moist. Drip irrigation provides a highly beneficial high moisture content in the soil where plants can thrive since water applications are made more frequently (typically every 1–3 days) than with other methods.

Drip irrigation benefits include disease prevention by reducing water contact with plant leaves, stems, and fruit. Because the system is so efficient, it saves time, money, and water.

It reduces the need for labor.

Major components of a drip irrigation system are as follows:

- Pump station.
- Bypass assembly.
- Control valves.
- Filtration system.
- Fertilizer tank/Venturi.
- Pressure gauge.
- Mains/submains.
- Laterals.
- Emitting devices.
- Micro tubes (Fig. 6.5).

Fig. 6.5 Drip irrigation system in India. (https://www.syngenta.co.in/drip-irrigation-support)

6.9.4   Sprinkler Irrigation

A sprinkler irrigation system utilizes a pump to apply water under high pressure. It is a modelled series of rainfall with adjustable frequency, time, intensity, and drop size range. Sprinkler systems are intended to distribute water to the field without relying on the soil surface for water distribution, in contrast to surface irrigation.

Due to the wide range of discharge capacities, water is disseminated through a network of pipes, sprayed into the air, and irrigates most soil types.

It is a modelled series of rainfalls with adjustable frequency, time, intensity, and drop size range. Sprinkler systems are intended to distribute water to the field without relying on the soil surface for water distribution, in contrast to surface irrigation.

Sprinklers are designed and organized to apply water at rates less than the rate of soil infiltration to prevent water logging and surface run off. These systems are employed in agricultural and horticultural production, as well as in landscape and lawn applications.

Some of the benefits include its suitability for all types of agricultural conditions (except heavy clay soil).

Water distribution that is uniform and efficient. Less land loss means more land available for farming. Water loss is minimal.

Some of its drawbacks include a high initial capital cost, extensive maintenance requirements, and high operating pressure.

- The major components of sprinkler irrigation systems are as follows:
- Pumping station or header assembly.
- Bypass valve.
- Fertilizer tank.
- Filtration system.

Fig. 6.6 Sprinkler irrigation system

- Pressure gauges.
- Control valves.
- HDPE/PVC pipes.
- QRC pump connector.
- Sprinkler nozzles.
- Service saddle (Fig. 6.6).

6.10 Fertigation

Fertigation is the administration of plant nutrients by irrigation. Chemical fertilizers are dissolved in water in the irrigation system, leading to delivery of water and fertilizer to the plant simultaneously conserving both costly water and fertilizer inputs. Hagin et al. (2002) suggest that fertigation is a cutting-edge agro-technique that offers a fantastic potential to enhance productivity and reduce environmental damage.

Fertilizer application time, volume, and concentration can all be readily managed via fertigation. Open/wild irrigation coupled with fertilizer application can lead to wastage of the latter along with the uneven distribution in the field. This directly or indirectly increases the cost of production for the farmer. Fertigation coupled with pressurize irrigation systems like drip, sprinkler, and micro sprinklers solves this problem.

6.10.1   A Brief History of Fertigation

The first known instance of fertigation occurred in ancient Athens (400 BC), when tree groves were watered using sewage from the city. Commercial fertigation began in the middle of the twentieth century. Although liquid ammonia was likely the first liquid fertilizer to be manufactured commercially, ammonia is no longer a significant source of nitrogen in the current fertigation scene.

6.10.2   Need for Fertigation

The need for fertigation is explained in the below mentioned points:

1. Variations in fertilizer usage among states and crops due to uneven development in fertilizer use.
2. Rapid extraction of nutrients from the soil (depletion of soil fertility as a result of insufficient and uneven fertilizer usage).
3. A decline in fertilizer response from crops.
4. The output of fertilizer industry has stagnated.
5. Growing reliance on imported fertilizer (mostly for P and K).
6. A weakening link between the usage of fertilizers and the production of food grains.

6.10.3   Advantages of Fertigation

- It eliminates the need for manual labor, hence cutting down costs of production.
- It reduces the wastage of nutrients as well as water coupled with high nutrient use efficiency.
- Fertigation leads to uniformity in fertilizer application.
- There is no effect of undulating terrain in the case of fertigation.

6.10.4   Disadvantages of Fertigation

- Chemical blockage can result from fertilizer interactions with calcium, magnesium, and bicarbonates in water.
- Equipment for fertigation that is corrosion resistant is required.
- It is only suitable for liquid or fertilizers that are easily soluble.
- Some micronutrients and phosphoric fertilizer may precipitate in a micro-irrigation system.

6.11   Problems in Water Management

Water is an indispensable part of agricultural and animal husbandry. It is high time for people to understand how to use and manage water resources properly, especially when it comes to agriculture. Water management is one of the most important concerns that humanity is now facing. Recent decades have brought this situation's seriousness to light. There are several water-related challenges in agriculture, including monitoring water pollution, reusing water, maintaining irrigation water pipeline distribution systems, and providing drinking water for livestock. Competition for water resources is also anticipated to rise as a result of population expansion, urbanization, and climate change, with an emphasis on agriculture.

Inadequate policies, severe institutional underperformance, and funding constraints are frequently barriers to improving the water management in agriculture (World Bank, 2022).

The good news is that, despite the abovementioned limitations, the agricultural water management industry is actively reinventing itself to provide contemporary and sustainable services while collaborating with local farmers. It suggests a unique method to managing risks associated with larger social and economic water-related repercussions while also constructing resilient water services and maintaining water supplies. This may be accomplished by improving incentives for innovation, reforms, and accountability. It also supports the management of watersheds and the greening of the industry.

6.12   Water Use Efficiency of Crops

The ability of a cropping system to turn water into plant biomass or grain is known as water use efficiency (WUE). It takes into account both rainfall throughout the growing season and the consumption of water that has been stored in the soil.

The soil's capacity to hold water, the crop's availability to that water and seasonal rainfall, the ability of the crop to transform water into biomass, and the capacity of the crop to transform biomass into grain (harvest index) are all necessary processes for efficient use of water by a crop.

The factors that affect WUE include nature of the plant, climatic conditions, soil moisture content, fertilizers, and plant population.

Mathematically, the yield of a marketable crop produced per unit of water utilized for evapotranspiration is known as WUE. It is written as

$$WUE = Y / ET$$

Where

Y is the marketable yield (kg/ha).
ET is evapotranspiration.
WUE is water usage efficiency (Kg/ha-mm) (mm).

6.13   Concept of Drainage

Drainage in agriculture is the process of removing extra water from the soil's surface or below, sometimes referred to as free water or gravity water, in order to improve the soil's conditions for plant development. In order to boost plant development in some circumstances or lessen the buildup of salt in the root zone of the crop, drainage also involves lowering the groundwater table (GWT) below the root zone.

Both drainage and irrigation are essential for a good crop yield. Drainage and irrigation must work together and cannot be isolated from one another. Excessive soil moisture can significantly damage crops and the soil, even for a little period of time.

While drainage is crucial for preventing excess moisture in the root zone, irrigation supplies the soil with the moisture it needs for plants to develop to their full potential. The excessive moisture may have its source in ditches, over-irrigation, canal or reservoir seepage, or excessive rainfall.

6.13.1   Types of Drainage

There are majorly two types of drainage: (a) surface drainage and (b) subsurface drainage.

Surface Drainage: Surface drainage is the process of removing extra water from the surface of the land. For quick water disposal in locations with excessive rainfall, adequate surface drainage must be set up. Surface water sources include things like rainfall, snowmelt, sewage runoff, seepage from neighboring higher terrain, etc.

Subsurface Drainage: These covered drains are buried beneath the soil's surface. They do not obstruct the usual operation of farm implements or cultivation techniques, and no space is lost in their construction. Subsurface drains are built using many sorts of materials. Short clay, concrete, or plastic pipes, fibrous wood products, covered stone drains, and bituminous fibrous materials can all be examples of these.

6.13.2   Objectives of Drainage

- Draining the soil of extra water.
- Betterment of soil aeration.
- To prevent nutrient loss from plants.
- To stop soil erosion.
- To stop salt from building up on the soil's surface.

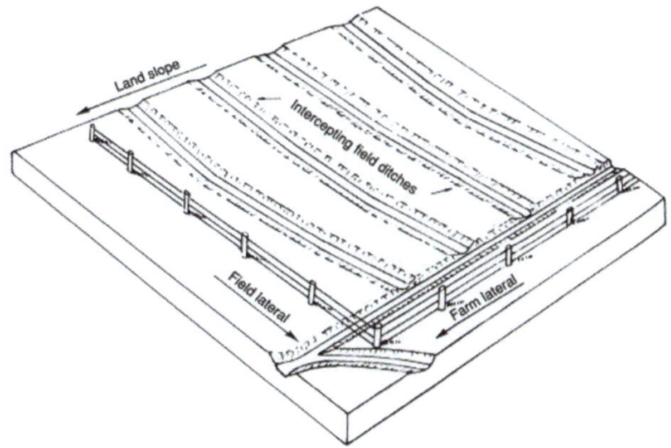

Fig. 6.7   Surface drainage. (Pavelis, 1987)

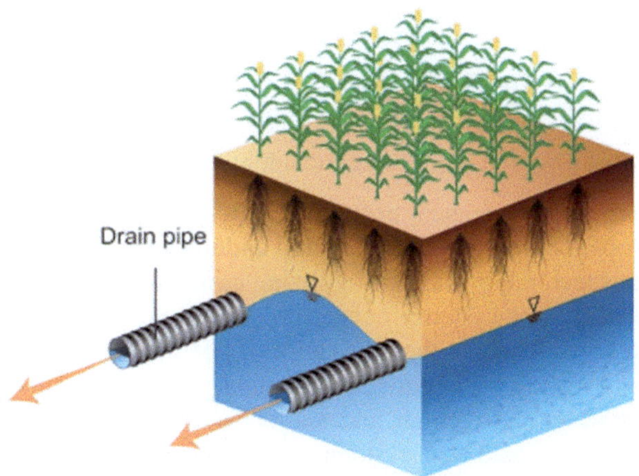

Fig. 6.8   Subsurface drainage (E. Ghane, MSU.)

- To create conditions that will promote the growth of plant roots (Figs. 6.7 and 6.8).

6.14   Introduction to Rainfed Agriculture and Watershed Management

In India, agriculture is the main source of income for roughly two thirds of the population. As ancient as agriculture itself is rainfed farming. Dryland agriculture refers to agricultural production solely in rainfed situations.

Agriculture that relies on rainfall of more than 1150 millimeters per year is known as rainfed agriculture. During the crop cultivation period, crops are not under stress from soil moisture because of rainfed practices. The elimination of extra water is frequently emphasized upon in the subject of rainfed agriculture. These humid areas have a growth season that lasts for more than 120 days.

Watershed is the region of land, which is responsible for draining or shedding water into a particular receiving water body, such as a lake or river. By regulating the use of those land and water resources in a holistic way, watershed management is the process of putting land use practices and water management practices into place to safeguard and enhance the quality of the water and other natural resources within a watershed.

Through a comprehensive approach, watershed management aims to stop land deterioration and maximize agricultural output. Watershed management entails the prudent use of land and water resources for maximum and sustained productivity with the least amount of risk to the environment.

6.15   Future Prospects of Rainfed Agriculture

Future prospects of rainfed agriculture include the following:

- Accurate agro-meteorological examination of the interaction between climate, soil, and crops for effective crop planning and management.
- Although this technology aids in stabilizing agricultural output in dry land locations, water collecting systems for crop production are unprofitable due to runoff and storage costs. There is therefore a lot of potential to stabilize output in dry land regions by using water collection methods on a community level.
- For effective water management and water conservation, the proper implementation of alternative land use systems, such as agroforestry, horticulture, and fodder-based farming systems, is necessary.
- Farmers who work on dry ground are still hampered by weather irregularities. A seed bank must be developed with government backing in order to assist farmers in implementing contingency strategies, as seed is the major input in the adoption of new farming technology.

6.15.1   Objectives of Watershed Management

- To prevent destructive runoff and soil erosion.
- To manage the watershed to reduce landslides, droughts, etc.
- Utilizing local natural resources to enhance agriculture and related industries can help the participants' socioeconomic situation.
- To save, enhance, and sustainably use the natural resources for production.

6.15.2   Droughts

Drought is a protracted period of dryness in the natural climatic cycle that can occur anywhere on the planet. It is a slow-onset calamity caused by a lack of precipitation, which results in a water deficit. Drought has major consequences for health, agriculture, economy, energy, and the environment. Eighty-two percent of the all the devastation caused by droughts is experienced by agriculture (FAO, 2021).

Climate change in recent years has shown to increase the global temperatures, thereby making already dry regions drier and previously rainy regions wetter. This implies that, in dry areas, increasing temperatures cause water to evaporate more quickly, which either increases the likelihood of drought or lengthens existing droughts. In the last 10 years, floods, droughts, tropical cyclones, heat waves, and severe storms have been responsible for between 80% and 90% of all known disasters caused by natural hazards (WHO, 2022).

Wilhite and Glantz1 classified drought definitions into four primary ways to measurement: meteorological, hydrological, agricultural, and socioeconomic. The first three techniques are concerned with methods of measuring drought as a physical occurrence. The last chapter examines drought in terms of supply and demand, following the repercussions of water scarcity as they ripple through socioeconomic systems (NDMC University of Nebraska).

- *Meteorological Drought.*

- Meteorological drought is often described by the quantity of dryness (in relation to some "normal" or average amount) and the length of the dry spell. Definitions of meteorological drought must be regarded region specific since atmospheric conditions that result in precipitation deficits vary greatly from region to region.
- *Agricultural Drought.*
- Agricultural drought connects different meteorological (or hydrological) drought features to agricultural consequences, concentrating on precipitation shortages, discrepancies between actual and potential evapotranspiration (PET), soil water deficits, lower groundwater or reservoir levels, and so on.
- *Hydrological Drought.*
- The consequences of precipitation (including snowfall) deficits on surface or subsurface water supplies are referred to as hydrological drought (i.e., streamflow, reservoir and lake levels, groundwater). Watershed or river basin scales are frequently used to characterize the frequency and severity of hydrological drought. Although all droughts are caused by a lack of precipitation, hydrologists are more interested with how this lack of precipitation manifests itself across the hydrologic system. Hydrological droughts typically occur out of sync with or after meteorological and agricultural droughts.
- *Socioeconomic Drought.*
- Socioeconomic definitions of drought combine meteorological, hydrological, and agricultural drought factors. It varies from the other forms of drought in that its occurrence is determined by the time and spatial processes of supply and

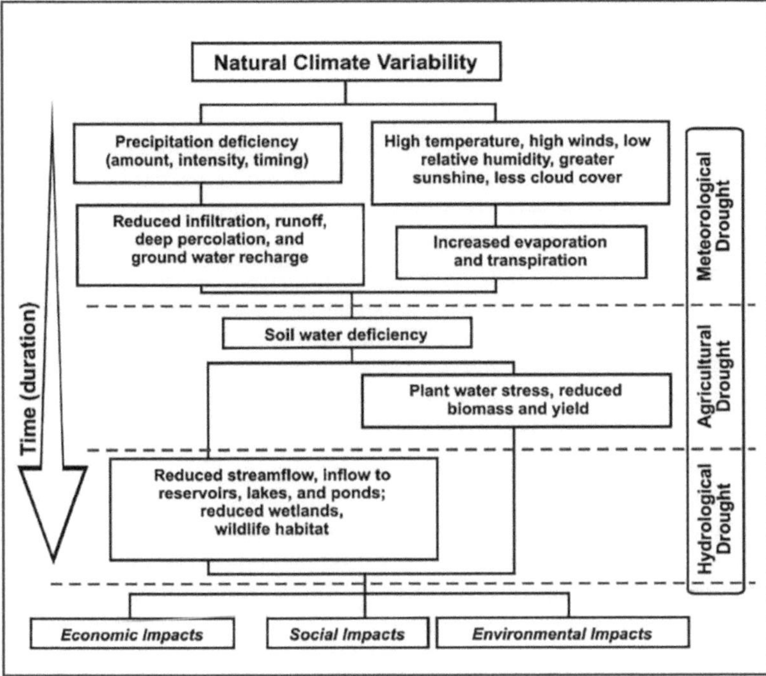

Fig. 6.9 Sequence of drought occurrence and impacts for commonly accepted droughts (NDMC)

demand to define or categorize droughts. When demand for an economic commodity exceeds supply due to a weather-related deficiency in water supplies, this is referred to as socioeconomic drought (Fig. 6.9).

- *Ecological Drought*: A more recent initiative focuses on ecological drought, which is described as "a sustained and widespread deficiency in naturally accessible water resources—including changes in natural and regulated hydrology—that creates multiple stresses across ecosystems."

6.15.3 Floods

A flood is defined as an excess of water on land. A river may acquire more water from strong rains or other natural calamities. When this occurs, the water overflows from its typical course in the river bed and onto dry ground. Floods are frequently brought on by prolonged periods of heavy rain, quick snowmelt, or storm surges from tropical cyclones or tsunamis in coastal regions.

Floods wreak havoc across a large area, killing people, devastating all the agriculture in the area, damaging private property, and destroying vital public health facilities. More than 2 billion people worldwide were impacted by floods between 1998 and 2017 (WHO).

There are various types of floods. They are mentioned below:

* *Coastal Flooding.*
  Severe storms frequently cause the most damage in coastal locations, especially if they have intensified over the oceans. Extreme weather conditions and high tides can raise sea levels, which can occasionally lead to coastal flooding. Coastal flooding is anticipated to become a frequent and seriously problematic issue as global warming advances.
* *River Flooding.*
  River flooding is one of the most prevalent kinds of inland flooding, occurring when the capacity of a body of water is exceeded. Rivers require effective flood defenses, particularly in flat or densely populated areas.
* *Flash Flooding.*
  Flash flooding occurs when the earth cannot absorb the water as rapidly as it falls due to high and abrupt rainfall. Even while these floods often pass quickly, they may be hazardous and move swiftly when they do. By using effective drainage systems and preventing overdevelopment on floodplains, flash flooding can be averted.
* *Groundwater Flood.*
  In contrast to flash floods, groundwater flooding happens gradually. The earth becomes so saturated with water from prolonged rain that it can no longer absorb any more. Flooding results from water rising over the ground's surface in this situation. Flooding of this kind can linger for weeks or even months at a time.
* *Drain and Sewer Flooding.*
  The weather is not always to blame for sewer flooding. They could happen due to rain as well as a blockage or other drainage system malfunction. Internal (inside a structure) or exterior flooding of drains and sewers is also possible.

6.16 Water Harvesting

A technique for gathering and storing water for later use is called water harvesting. Direct rainwater collection is the usual for water harvesting. Lakes, groundwater, and rivers are secondary water sources, whereas rainfall is the major supply. Instead of letting the rainwater to flow off, rainwater harvesting involves collecting it and storing it for later use. On the other hand, it is employed for irrigation, residential use when properly treated, rearing animals, orchards, and gardens. The water that is collected can also be utilized for groundwater recharging, longer-term storage, and consumption.

Water harvesting is typically used for agriculture in arid and semiarid countries, and it works best in places close to hillsides or when cultivation is challenging owing to a significant amount of bare soil.

Ridges of soil are built to catch rainfall and stop it from pouring down hills and slopes, improving irrigation in dry areas. It is possible to collect water from rooftops and build dams and ponds to store vast amounts of rainwater, and all of these methods may ensure that there is enough water on hand to irrigate growing crops even when there is little or no precipitation.

Farmers can store water through water harvesting and then release it when it is increasingly scarce.

6.16.1   Advantages of Water Harvesting

- It decreases soil erosion, storm water runoff, flooding, and surface water pollution caused by fertilizers, pesticides, metals, and other sediments.
- It is an ideal source of water for landscape irrigation because it contains no chemicals or dissolved salts and is mineral-free.
- It can be used to irrigate gardens at home as well as crops on a commercial scale.
- Water harvesting strategies are critical in arid and semiarid environments for dry land farming.

6.16.2   Disadvantages of Water Harvesting

- Water harvesting necessitates routine maintenance.
- Installation also demands some technical knowledge.
- Rainwater availability might be limited in conditions of minimal and no rainfall.
- If not properly installed, it may attract mosquitoes and other waterborne diseases.
- Storage limitations are one of the major downsides of the rainwater harvesting technology.
- Poorly made water jars and containers may develop algae growth and get infested with insects, reptiles, and rats.

6.16.3   Types of Water Harvesting

There are majorly two types of water harvesting:

(a) *Rooftop Rain Water Harvesting*: Rooftop harvesting involves collecting rainwater from a building or home's roof by turning it into a catchment. It can be either diverted to an artificial recharge system or kept in a tank.

Fig. 6.11 Rainwater harvesting in Nepal (2013)

(b) *Surface Runoff Harvesting*: By using the proper techniques, it is a technique whereby rainfall that is now flowing as surface runoff is captured and used to recharge aquifers (Fig. 6.11).

MCQs

1. What percent of water of total water on earth is suitable for human use?

   - 2%
   - 3%
   - 6%
   - 7%

2. Irrigated agriculture accounts for _____ of the total cultivated land.

   - 20%
   - 30%
   - 40%
   - 25%

3. Eighty-two percent of the all the devastation caused by droughts is experienced by.

   - Infrastructure.
   - Agriculture.
   - Human health.
   - Horticulture.

4. Which word mentioned below refers to, "the rate of evapotranspiration from a wide surface of 8–15 cm tall green grass cover of uniform height that is actively growing, fully shading the ground, and not water-stressed."

   - Potential Evapotranspiration (PET).
   - Reference Evapotranspiration (ET_O).
   - Actual Evapotranspiration.
   - None of the above.

5. The amount of water consumed in a 24-hour period is referred to as.

   - Peak Consumptive Use.
   - Daily Consumptive Use.
   - Hourly Consumptive Use.
   - None of the above.

6. About ____ of the water absorbed by plants is used for metabolic processes and body weight production.

   - 8%
   - 15%
   - 2%
   - 5%

7. Submergence under water at tillering and flowering stages are the critical stages of crop growth for irrigation of?

   - Wheat.
   - Maize.
   - Rice.
   - Barley.

8. Early vegetative growth, flowering, and pod development are critical phases for?

   - Pulses and Legumes.
   - Maize.
   - Rice.
   - Barley.

9. Commercial fertigation began in the middle of the ________.

   - Nineteenth Century.
   - Twentieth Century.
   - Eighteenth Century.
   - None of the above.

10. Wheat has a water requirement of (in mm)?

- 500
- 1800
- 280
- 1200

References

FAO. (1998). *Production yearbook* (Vol. 52). Food and Agricultural Organization of the United Nations.

FAO. (2021). *AQUASTAT database*. Food and Agricultural Organization of the United Nations.

FAOSTAT. (1999). *FAOSTAT statistical database*. Food and Agricultural Organization of the United Nations. http://apps.fao.org

Hagin, et al. (2002). *Fertigation—Fertilization through irrigation* (p. 23). IPI Research.

Kabiri, R. (2010). *Effect of salicylic acid to reduce the oxidative stress caused by drought in the hydroponic cultivation of Nigella sativa (Nigella sativa).*

Kumar, R., Singh, R., & Sharma, K. (2005). Water resources of India. *Current Science, 89*, 794–811.

Pavelis GA (ed.) (1987) *Farm drainage in the United States: History, status, and prospects.* Miscellaneous publication no. 1455. Economic Research Service. US Department of Agriculture. US Government Printing Office (Surface Drainage Diagram).

Plane mad—Own work International borders: University of Texas map library—India Political map *2001* Disputed borders: University of Texas map library—China-India Borders—Eastern Sector *1988* & Western Sector *1988*—Kashmir Region 2004—Kashmir Maps. State and district boundaries: Census of India—*2001* Census State Maps—Survey of India Maps. Other sources: US Army Map Service, Survey of India Map Explorer, Columbia University Map specific sources: Rainfall map., CC BY-SA 3.0, https://commons.wikimedia.org/w/index.php?curid=1213463.

Rahmani, N. (2006). Effect of irrigation and nitrogen application on the quantity and quality of medicinal plant Marigold (Calendula Officinalis L.). University of Takestan Branch. Responses of plants to drought stress. *Journal of Agricultural Research, 6*(9), 2026–2032.

Rajanna, G. A., Dass, A., & Venkatesh, P. (2019). Excess water stress: Effects on crop and soil, and mitigation strategies. *International Journal of Agronomy, 6*, 48–53.

Reddy, S., & Reddy, S. (2004). *Sustainable agriculture and development*. Springer.

Singh, V., Patel, R., Kumar, S., Ahirwal, A., & Sasode, D. (2021). Moisture stress and their effect on crops 2. https://www.researchgate.net/publication/350439278_MOISTURE_STRESS_AND_THEIR_EFFECT_ON_CROPS

Sprinklers in vineyard. —Trentino-Alto Adige, Italy. Marek Ślusarczyk. https://commons.wikimedia.org/wiki/File:10_Sprinklers_in_vineyard_-_Trentino-Alto_Adige,_Italy.jpg

Todorovic, M. (2005). Crop water requirements. In J. H. Lehr & J. Keeley (Eds.), *Water encyclopedia: Surface and agricultural water, AW-59* (pp. 557–558).

UN. (1997). *Commission on sustainable development. Comprehensive assessment of the freshwater resources of the world*. Report of the Secretary General, 39p. https://www.worldbank.org/en/topic/water-in-agriculture

Vince Ördög. (2011). *Plant Physiology 10*. https://www.esalq.usp.br/lepse/imgs/conteudo/Plant-Physiology-by-Vince-Ordog.pdf

World Bank. (2022). *Global economic prospects*. World Bank Group. https://www.worldbank.org/global-economicprospects

World Health Organization. (2022). *Global report on health equity*. World Health Organization. https://www.who.int/publications/global-report-health-equity

Chapter 7
Precision Agriculture

Abstract This chapter introduces the concept of precision agriculture (PA). PA is a concept of using new technologies and information collected from the fields to optimize the field-level management with regard to crops, environmental protection, and economy. The underlying principle of PA is doing the right thing, in the right place, at the right time. This chapter explores the need, requirement, aim, and processes of precision agriculture along with its merits and demerits. Toward the end of the chapter, there is a shift of focus toward more modern form of precision agriculture with reference to modern-day tools and technologies.

Keywords Mapping · Precision farming · VRT · Drones · Robotics yield mapping

7.1 Introduction

In simple terms, everything that makes crop cultivation and animal husbandry more precise and under control is referred to as "precision farming." The application of artificial intelligence, a variety of electronics, and communication technologies is a crucial part of this farm management strategy. Precision agriculture heavily relies on information technology for its proper functioning.

Precision agriculture is the adoption of the newest and most cutting-edge approaches, concepts, techniques, and technology to replace traditional agricultural practices and make them sustainable.

Further precision agriculture can be defined as follows:

- Precision agriculture or satellite farming or site-specific crop management is a farming based on observing, measuring, and responding to inter- and intra-field variability in crops.

Or

Fig. 7.1 Precision agriculture technologies (AGRIVI)

- PA is an information- and technology-based farm management system to identity, analyze, and manage variability within fields for optimum profitability, sustainability, and protection of land resources.

  Or

- Precision application of technologies and input based on soil, crop weather, and market demands to maximize sustainable productivity and profitability.

  Or

- Precision forming is generally defined as an information and technology best for management system to identify, analyze, and manage variability within fields for optimum profitability, sustainability, and production of land resource (Fig. 7.1).

7.2 Need for Precision Agriculture

- Increased agricultural output closely corresponds to the demand for precision agriculture, particularly in India where the majority of farmer landholdings (86%) are small and marginal (Agricultural Census 2015–2016).
- Moreover, precision agriculture has shown significant reduction in soil degradation.
- In PA, there has been a dramatic decrease in the usage of chemicals in crop production as a result of the extensive use of cutting-edge technologies to predetermine any sort of infestation in the standing crop.
- PA also improves the efficient utilization of water resources.
- By promoting modern farming techniques, PA aids in raising the quality, quantity, and profitability of agricultural output, which in turn raises farmers' social standing in society.

7.3 Aim of Precision Agriculture

1. Crop science by more closely matching crop practices to crop demands (e.g., fertilizer inputs)
2. Environmental protection: through lowering environmental hazards and farming's environmental footprint (e.g., limiting leaching of nitrogen)
3. Economics: by making businesses more competitive through more effective methods (better input management, such as the use of fertilizer)

7.4 Advantages of Precision Agriculture

Along with the abovementioned benefits of precision agriculture, there are various other merits of adopting PA, and they are mentioned below:

- GPS makes it simple to survey agricultural areas. Additionally, maps of soil traits and yield are also possible.
- PA includes dissemination of knowledge about agricultural techniques to increase crop quality, quantity, and production at a lower cost.
- By optimizing agrochemical goods, it will reduce environmental risk, especially with regard to nitrate leaching and groundwater contamination.
- Nonuniform fields can be split into more manageable plots according to their particular needs.
- It offers chances for improved resource management, which minimizes the waste of all resources.

7.5 Disadvantages of Precision Agriculture

- Farmers may be discouraged from using this farming technique by its high capital costs.
- Before being used, precision agriculture techniques still need professional guidance as they are in the early stages of development.
- Before the system has collected enough data to be completely implemented, it may take several years.
- It is a very challenging job, especially the data collection and analysis.
- Moreover, not every crop can benefit entirely from precision farming. Before deploying this technology, farmers must first take on a number of technological, scientific, mechanical, and financial challenges (Fig. 7.2).

Fig. 7.2 Precision agriculture: a comprehensive approach. (Banu, 2015)

7.6 History of Precision Agriculture

The first workshop was conducted in Minneapolis in 1992, which is when PA first came into existence.

However, GPS was made available as a general-purpose device in the years 1970–1980. The first yield meter mounted on a combine harvester was created around the same period, and farmers benefited greatly from this invention. First yield maps were developed in the year 1984 (with GPS). Then, in 1991, application maps (based on the GIS) were released, and the first use of variable-rate technology was created.

The first precision agriculture training took place in Minneapolis in 1992. The invention of ground-based, satellite-based, and aerial sensing devices that assess crop status and the amount of chlorophyll emerged between 1995 and 1998.

We witnessed the introduction of soil electrical conductivity measurements and aerial/satellite images to assess crop condition between 1999 and 2002. RTK devices were created and used in agriculture in the year 2000. Scientists first experimented with weed detection systems and accurate seeding systems from 2000 to 2002. Agriculture started using auto-steering in 2003. The first controlled traffic systems among farms were put into place in 2008. The development of UAVs (drones) for application maps occurred in the same year. The first robotic technologies in high value crops and horticulture were introduced in 2015.

7.7   Components of Precision Agriculture/Tools of Precision Agriculture

Researchers found it challenging to link agricultural yields and production methods to the variability of resource availability in the past. In order to maximize crop production and reduce input costs with the least amount of harm to the environment, soil, water, and human health, precision farming techniques that are specific to a given area are implemented. Geographical information systems (GIS), geographic positioning systems (GPS), remote sensing (RS), variable rate technology (NDVI), and yield monitoring are the primary components of precision agriculture. Some of them are discussed in the below mentioned paragraphs.

7.7.1   Geographical Information Systems (GIS)

First used in 1960, the geographic information systems (GIS) are a combination of computer hardware and software that generates maps by combining feature characteristics and location data. The GIS can display geographic position data, allowing for a visual standpoint for assessment. GIS is a tool for gathering, storing, retrieving, and analyzing information about the characteristics of a specific location.

An agricultural GIS's ability to hold layers of data is a key feature. Each layer, or "coverage," is made up of topologically related geographic characteristics like surface drainage, soil types, remotely sensed data, and soil survey maps. The development of a model to link different soil, environmental, and crop parameters to crop yield uses GIS as a very effective aid.

7.7.2   Global Positioning System (GPS)

It is defined as a navigation system comprised of a network of satellites that enables the identification of locations in the field within a meter of an actual site (100 m to 0.01 m) (Hakkim et al., 2016).

GPS allows for precise mapping of farms and, in conjunction with appropriate software, informs the farmer about the status of his crop and which parts of the farm require what input such as water, fertilizer, and/or pesticides, or if there is any type of obstruction, pest occurrence, or weed invasion.

This information is provided in real time, which means that continuous position information is provided while in motion. Measurements of soil and crop parameters can be mapped when precise location data is available in real time with the help of GPS.

7.7.3   Remote Sensing

The art and science of collecting information about real-world objects or locations without coming into close proximity to them is known as remote sensing. The principle underlying remote sensing is the assessment of the earth's features using electromagnetic spectrum (e.g., visible, infrared, and microwaves). Data sensors, which take up the data for us, can be handheld devices, aircraft-mounted, or satellite-based.

Remote sensing data can be used to assess crop health. Plant stress caused by moisture, nutrients, pest infestations, crop diseases, and other plant health issues are easily identifiable via aerial images.

Remote sensing is an important tool in precision agriculture that has shown significant benefits when combined with GIS, GPS, satellites, and other technologies. Some of the ways remote sensing utilizes the upsides of precision agriculture are as follows (Bharteey et al., 2020):

- Soil mapping, determining climate and land characteristics, detecting crop nutrient deficiency, and conducting vegetative analysis
- Crop yield estimation and production forecasting in India
- Satellite-based agro-advisory service
- Pest control
- Evaluation and monitoring
- Evaluation of soil site suitability
- Estimation of soil moisture
- Assessment and monitoring of floods (Fig. 7.3)

7.7.4   Variable Rate Technology (VRT)

Fertilizer, pesticides, lime, irrigation water, and other agricultural inputs can all be applied to an area at various rates thanks to variable-rate technology (VRT).

Simple flow rate control to more complex management of rate, chemicals, and application pattern are all examples of variable-rate application (VRA). Farm level

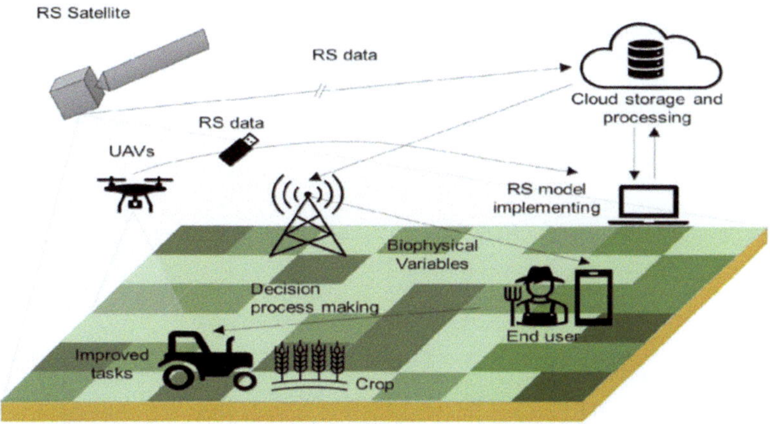

Fig. 7.3 Remote sensing in agriculture. (Erazo-Mesa et al., 2022)

controllers for the application of VRT fertilizer and herbicide have lately become available. VRT irrigation devices and VRT application of animal manure are additional advancements. There are three parts to the variable rate applicator. These include actuators, locators, and control computers.

VRT comes in two different forms:

1. *Map-based control*: Prior to the procedure, a map of application rates for the field is created.
2. *Real-time control*: Using data gathered throughout the operation, decisions are made about what rates to follow in various locations.

7.7.5 Precision Irrigation System

In order to effectively manage the water variability in the field and increase crop productivity and water use efficiency while lowering energy costs associated with irrigation, precision irrigation entails applying the correct amount, timing, and quality of water to the crop.

One of the crucial instruments used in precision agriculture is a sprinkler irrigation system with GPS-based controllers. These irrigation systems use wireless sensor networks to watch the soil and surrounding conditions as well as the operational parameters, such as frost, pressure levels, flow etc., of the irrigation system (Fig. 7.4).

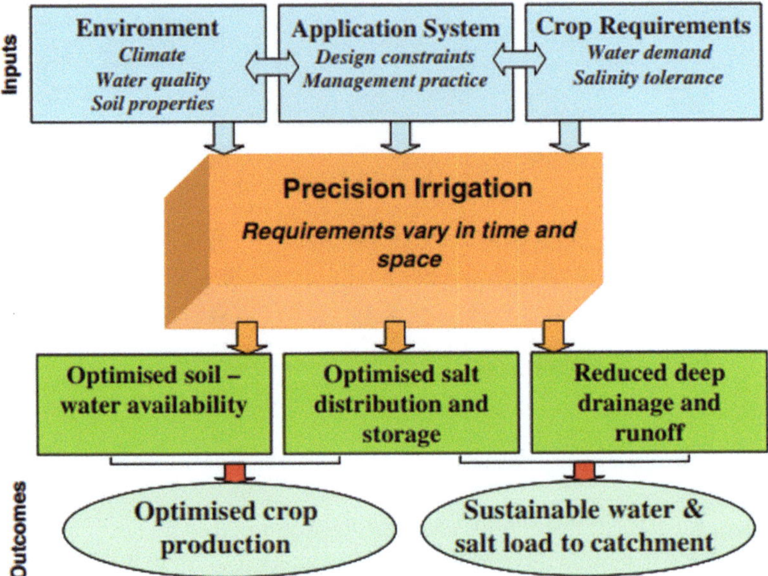

Fig. 7.4 Inputs and outputs of a precision irrigation system. (Raine Steven et al., 2005)

7.7.6   Drones and Use of Robotics

Drones are adaptable and have applications across many industries, including agriculture. Drones' use in precision agriculture has greatly increased since artificial intelligence (AI) was integrated into them because it is practical for farmers to use. Drones are useful for a number of tasks, including data collection, disease detection, crop and field monitoring, water, fertilizer, pesticide dissemination, and so on.

Different drone types are being investigated to determine which has the most promise for agriculture. Quadcopters and other multirotor drones are the finest option for fertilizing crops. Fixed-wing drones are perfect for fertilizing crops, but they are difficult to use because of their large structure and need for a large landing space.

One of the crucial areas that requires automation and smart devices that can carry out tasks that once required human involvement is the precision agriculture sector. Examples include sophisticated tractors, which are simply AI-based devices that have multiple technologies like sensors, radars, and GPS systems to carry out the tasks without the need for an operator (Kharkovyna, 2019).

7.7.7 Yield Monitoring and Mapping

One of the most crucial things a farmer must monitor is yield mapping and monitoring, and scientists have developed a method for doing so accurately. The main portion of the needed information, particularly grain yield, its moisture levels, etc., is gathered with the aid of monitors that are attached to farming equipment, such as tractors or combine harvesters. Grain flow in the clean-grain elevator of a combine is constantly measured and recorded by grain yield monitors in highly automated systems. When used correctly, this data offers significant feedback for assessing the impacts of managed inputs like seed, pesticides, fertilizer amendments, and cultural practices like irrigation and tillage.

Given the potential for weather to have a significant impact on yield measurements from a single year, it is always advisable to look at yield data from more than 10 years, including data from years with extreme weather. This helps to determine whether the observed yields are caused by management or climate-induced. For metadata and the analysis of spatial and temporal trends, this method delivers meaningful, instructive, and factual information. GPS and a yield monitoring system can also be used for yield mapping (Bakhtiari & Hematian, 2013).

MCQs

1. Majority of farmer landholdings by farmers in India are:

   - Large
   - Humongous
   - Small and Marginal
   - Medium

2. What percentage of majority of farmer landholdings by farmers in India are small and marginal:

   - 74%
   - 86%
   - 66%
   - 91%

3. The first workshop for precision farming was conducted in Minneapolis in which year?

   - 1992
   - 1998
   - 2002
   - 1988

4. The first robotic technologies in high value crops and horticulture were introduced in

   - 2016
   - 2014
   - 2015
   - 2020

5. Which tool is used for gathering, storing, retrieving, and analyzing information about the characteristics of a specific location.

   - GIS
   - GPS
   - Remote Sensing
   - VRT

6. Which technology enables the identification of locations in the field within a meter of an actual site (100 m to 0.01 m)

   - GIS
   - GPS
   - Remote Sensing
   - Yield Mapping

7. Quadcopters and other multirotor drones are the finest option for.

   - Yield Mapping
   - Pest Management
   - Fertilizer Application
   - All of the above

8. First yield maps were developed in the year _____

   - 1999
   - 1985
   - 1984
   - 1987

9. _______ data can be used to assess crop health. Plant stress caused by moisture, nutrients, pest infestations, crop diseases, and other plant health issues are easily identifiable via aerial images.

   - GPS
   - VRT
   - RS
   - All of the above

10. To hold layers of data is a key feature of?

   - GPS
   - VRT
   - Agricultural GIS
   - RS

References

Bakhtiari, A. A., & Hematian, A. (2013). Precision farming technology, opportunities and difficulty. *International Journal for Science and Emerging Technologies with Latest Trends, 5*(1), 1–14.

Banu, S. (2015). Precision agriculture: Tomorrow's technology for today's farmer. *Journal of Food Processing & Technology, 6*, 468. https://doi.org/10.4172/2157-7110.1000468

Bharteey, P. K., Deka, B., Dutta, M., Parit, R. K., & Maurya, P. (2020). Remote sensing application in precision agriculture: A review. *International Journal of Multidisciplinary Research and Development, 6*(11), 233–241.

Erazo-Mesa, E., Echeverri, A., & Gil, J. G. (2022). Advances in Hass avocado irrigation scheduling under digital agriculture approach. *Revista Colombiana de Ciencias Hortícolas., 16.* https://doi.org/10.17584/rcch.2022v16i1.13456

Hakkim, V., Joseph, E., Gokul, A., & Mufeedha, K. (2016). Precision farming: The future of Indian agriculture. *Journal of Applied Biology and Biotechnology, 4*(06), 68–72. https://doi.org/10.7324/jabb.2016.40609

Kharkovyna, O. (2019). *7 reasons why machine learning is a game changer for agriculture.* https://www.linkedin.com/pulse/7-reasons-why-machine-learning-game-changer-kharkovina-alexey

Raine, S., Meyer, W., Rassam, D., & Hutson, J. L. (2005). *Soil-water and salt movement associated with precision irrigation systems – Research investment opportunities* (p. 91). Land and Water Australia.

Chapter 8
Remote
Sensing and Geo- and Agri-Informatics

Abstract The transformative synergy of remote sensing, geoinformatics, and agricultural informatics in modern agriculture improves farmer convenience by using real-time data and maps to make better decisions, manage resources more efficiently, and boost crop yields. Remote sensing provides valuable insights through satellite and aerial imagery, capturing dynamic information on crop health, land use, and environmental conditions. Geoinformatics processes and analyzes this spatial data, creating detailed maps and models that enhance our understanding of agricultural landscapes. The integration of agricultural informatics complements these technologies by incorporating real-time data on crop management, weather patterns, and soil conditions. The profound impact of merging remote sensing, geoinformatics, and agricultural informatics ushers in a new era of intelligent and sustainable agriculture. This convergence empowers stakeholders with actionable insights, promoting efficiency, resilience, and environmentally conscious practices in modern farming.

Keywords RS · GIS · GPS and precision farming

8.1 Introduction Remote Sensing

Remote sensing is the acquisition of physical data of an object without touch or contact (Lintz & Simonett, 1976). Remote sensing is the science and technology of acquiring information about objects or areas from a distance without direct contact. This is typically done using sensors on aircraft or satellites, and the acquired data can be used for various purposes, including environmental monitoring, land-use planning, agriculture, forestry, disaster management, and more. Remote sensing includes all methods of obtaining pictures and other forms of electromagnetic records of the Earth's surface from a distance and the treatment and processing of the picture data. Remote sensing in widest sense is concerned with detecting and recording electromagnetic radiations from the target areas in the field of view of the sensor instrument. Remote sensing data are a major source of data for the mapping of resources like geology, forestry, water resources, land use, and land cover.

L. Ahmad et al., *Fundamentals and Applications of Crop and Climate Science*,
https://doi.org/10.1007/978-3-031-61459-0_8

Integration of the two technologies, remote sensing and GIS, can be used to develop decision support systems for a planner or decision-maker. Remotely sensed images can be used for two purposes, as a source of spatial data within GIS and using the functionality of GIS in processing remotely sensed data in both pictorial and digital modes. Images derived from optical and digital remote sensing systems mounted in aircraft and satellites provide much spatial information and major data as an input to GIS. Remote sensing data basically consists of wavelength intensity information acquired by collecting the electromagnetic radiation leaving the object at specific wavelength and measuring its intensity. Spatial patterns evident in remotely sensed images are interpreted in terms of geographical variation in the nature of material forming the surface of the earth. Such materials may be vegetation, exposed soil and rock, or water. These materials are not themselves detected directly by remote sensing, and their nature inferred from the measurements made. The characteristic of digital image data is that they can be adjusted so as to provide an estimate of physical measurements of properties of the targets, such as radiance or reflectivity (Mather & Koch, 2011).

8.1.1   Types of Remote Sensing

Broadly, there are two types of sensing systems to record the information about any target. They are active sensing system and passive sensing system. An active sensing system generates and uses its own energy to illuminate the target and records the reflected energy which carries the information content or entropy. Synthetic aperture radar (SAR) is one of the best examples of active sensing systems. These sensing systems operate in the microwave region of electromagnetic spectrum and include radiation with wavelengths longer than 1 mm. These systems do not rely on the detection of solar or terrestrial emissions as the solar irradiance in microwave region is negligible. The active remote sensing operation principles and the general details of latest imaging radar systems are described in the following chapter. The second type of remote sensing systems is passive systems mainly depending on the solar radiation operates in visible and infrared region of electromagnetic spectrum. The nature and properties of the target materials can be inferred from incident electromagnetic energy that is reflected, scattered, or emitted by these materials on the earth's surface and recorded by the passive sensor (e.g., a camera without flash). The remote sensing system that uses electromagnetic energy can be termed as electromagnetic remote sensing.

8.1.2   Components of Remote Sensing

The basic components of remote sensing include various elements and processes that contribute to the acquisition, transmission, and interpretation of data from a distance. Here are the fundamental components of remote sensing:

1. Energy Source or Illumination.

   Natural Source: The sun is the primary natural source of energy for remote sensing. Sunlight provides illumination across the electromagnetic spectrum. Artificial Source: Some remote sensing systems use artificial sources of energy, such as radar or lidar, which emit their own signals.

2. Radiation and the Atmosphere.

   Atmospheric Conditions: The Earth's atmosphere interacts with the incoming solar radiation and can affect the quality of data collected. Factors like clouds, aerosols, and gases can influence how much energy reaches the Earth's surface and how it is reflected or absorbed.

3. Interaction with the Target.

   The energy from the source interacts with the Earth's surface and features. Different surfaces (e.g., water, vegetation, and soil) interact with the energy in unique ways, influencing the type of data collected.

4. Recording of Energy by the Sensor.

   Sensor: Remote sensing instruments or sensors capture the energy reflected or emitted by the Earth's surface. Sensors can be mounted on satellites, aircraft, or ground-based platforms.
   Detectors: These are components within the sensor that measure the electromagnetic radiation in specific wavelengths. They convert the received energy into electrical signals.

5. Transmission, Reception, and Processing.

   Data Transmission: Collected data is transmitted from the sensor to ground stations or receiving systems. For satellite-based systems, this involves downlinking data to Earth.
   Data Reception: Ground stations receive and process the transmitted data.
   Data Processing: Raw data is processed to correct for atmospheric effects, geometric distortions, and other factors. This step prepares the data for interpretation.

6. Interpretation and Analysis.

   Image Interpretation: Analysts visually interpret remote sensing images to identify and classify features.

Digital Image Processing: Computer-based algorithms and techniques are applied to process and analyze digital images automatically.

7. Applications and Decision-Making.

The information derived from remote sensing data is used for various applications including land cover mapping, environmental monitoring, agriculture, urban planning, and disaster management. Decision-makers use the interpreted data to make informed choices and policies.

These components collectively enable remote sensing to be a valuable tool for understanding and monitoring the Earth's surface over large areas.

8.1.3   Use of Remote Sensing in Land and Water Resources

Remote sensing plays a crucial role in the monitoring and management of land and water resources by providing valuable information about their spatial distribution, condition, and changes over time. Here are some specific applications of remote sensing in the assessment and utilization of land and water resources:

8.1.4   Land Resources

1. Land Cover and Land Use Mapping.
   Remote sensing helps in classifying and mapping different land cover types, such as forests, agricultural fields, urban areas, and wetlands. Land use changes over time can be monitored, providing insights into urban expansion, deforestation, and agricultural practices.
2. Vegetation Monitoring and Health Assessment.
   Satellite imagery is used to monitor vegetation health, assess crop conditions, and detect diseases or stress factors affecting plant cover. This information is valuable for precision agriculture, optimizing irrigation, and predicting crop yields.
3. Forest Management.
   Remote sensing aids in assessing forest extent, monitoring deforestation, and estimating biomass. It supports forest management practices, such as identifying areas at risk of wildfires and planning for sustainable logging.
4. Natural Resource Exploration.
   Remote sensing data helps identify and locate natural resources such as minerals, oil, and gas through the analysis of surface features and geological characteristics.

5. Soil Mapping and Erosion Monitoring.

   Soil properties, types, and erosion patterns can be studied using remote sensing, assisting in land suitability assessments for agriculture and infrastructure development.

6. Land Degradation Assessment.

   Changes in land quality and degradation can be detected through the analysis of vegetation cover, soil moisture, and surface reflectance, helping in the formulation of conservation strategies.

8.1.5   Water Resources

1. Water Quality Monitoring.

   Remote sensing enables the assessment of water quality parameters such as turbidity, chlorophyll concentration, and pollutants in lakes, rivers, and coastal areas. This information is critical for managing water resources, especially in the context of drinking water supplies and aquatic ecosystem health.

2. Water Quantity Estimation.

   Remote sensing data, including satellite-based altimetry, can be used to estimate water levels in rivers, lakes, and reservoirs, aiding in water resource management and flood forecasting.

3. Wetland Mapping.

   Wetland ecosystems are identified and monitored using remote sensing to understand their extent, changes, and ecological functions. This is essential for conservation and habitat management.

4. Irrigation Management.

   Remote sensing helps optimize irrigation practices by assessing soil moisture levels, crop health, and water use efficiency, contributing to sustainable water resource management in agriculture.

5. Flood Monitoring and Prediction.

   Satellite imagery is used to monitor and predict floods by assessing changes in water levels and identifying areas prone to inundation. This information is crucial for disaster preparedness and response.

6. Snow and Glacial Monitoring.

   Remote sensing aids in monitoring snow cover and glaciers, providing insights into water availability and potential impacts on downstream water resources.

By leveraging remote sensing technologies, decision-makers and resource managers can make informed choices for sustainable land and water resource utilization, conservation, and environmental protection.

8.1.6   Introduction to Geography

Till a few decades back, the study of geography implied only a catalogue of names, and a student of geography felt quite satisfied if he could commit to memory such facts as the names of continents and oceans, the depths of seas and bays, the length of rivers, and the height of mountains. Geography was recognized as a description of the world and its inhabitants. But the geography of today is not merely a description or interpretation of the regions of the world. It is an enquiry, a study of the causes, an attempt to find out the why and how of all those facts and factors that go to influence the life of man on this planet. Modern geography studies earth as the home of man. It attempts to find out the relationship between man and his environment. It depicts the geographical conditions under which man lives and works. In other words, it studies man's adaptation to the environment in which he lives and works.

1. *Definition of Geography.*

    It is the study of the physical features of the earth and its atmosphere and of human activity as it affects and is affected by these, including the distribution of populations and resources and political and economic activities.

   (i) *Physical Geography.*

      Physical geography is the study of physical features of the earth. This includes study of landforms, drainage forms, water bodies and shorelines, minerals, soils, climate, natural vegetation, and life of animals.

   (ii) *Human Geography.*

      Human geography studies the man-made features of the earth. This includes cultural elements such as the density and pattern of population, the utilization of land, the types of and materials for the buildings, and the means of transportation and communication. Human geography is further divided into passive or static human geography: It deals with the action of nature upon man. Active or dynamic human geography: It deals with the reaction or action of man upon nature.

The factors that influence, determine, and shape the distribution and activities of humans include physical features, climate, vegetation, and animal life, determining the life and activities of humans in different areas of the world. Of these main factors, the physical features and the climate are the most important, and they exert the determining influence on the occupation and settlement of man in the different areas of the world. The type of vegetation in any region is controlled to a very great extent by the configuration of land and the climatic condition.

(iii) *Physical Features.*

      These include the configuration of land, its various formations, the drainage systems, and mineral content of the rocks and the soils. Human's habitats, occupations, mode of living, and capacity for work are primarily and directly influenced by the physical features of a particular area.

Climate is an extremely powerful factor so far as the occupancy of an area by human is concerned. The working capacity of human and his capacity to produce economic wealth is also determined by factors like temperature, rainfall, and winds, which are included in the climate of an area.

Vegetation: The vegetation depends upon both the physical features and the climate. The growth of plants and trees is determined by the texture of soil, the range of temperature, and the amount of moisture supply.

Animal Life: The animal life is determined directly and apparently by the vegetation cover but the primary influence is that of climate and physical features.

2. *Earth: Shape, Size, and Structure.*

Earth, with an average distance of 92,955,820 miles (149,597,890 km) from the sun, is the third planet and one of the most unique planets in the solar system. It formed around 4.5–4.6 billion years ago and is the only planet known to sustain life. This is because of factors like its atmospheric composition and physical properties such as the presence of water over 70.8% of the planet allow life to thrive. Earth is also unique however because it is the largest of the terrestrial planets (one that have a thin layer of rocks on the surface as opposed to those that are mostly made up of gases like Jupiter or Saturn) based on its mass, density, and diameter. Earth is also the fifth largest planet in the entire solar system. Earth's size as the largest of the terrestrial planets has an estimated mass of 5.9736×1024 kg. Its volume is also the largest of these planets at 108.321×1010 km^3. In addition, Earth is the densest of the terrestrial planets as it is made up of a crust, mantle, and core. The Earth's crust is the thinnest of these layers, while the mantle comprises 84% of Earth's volume and extends 1800 miles (2900 km) below the surface. What makes Earth the densest of these planets, however, is its core. It is the only terrestrial planet with a liquid outer core that surrounds a solid, dense inner core. Earth's average density is 5515×10 kg/m^3. Mars, the smallest of the terrestrial planets by density, is only around 70% as dense as Earth. Earth is classified as the largest of the terrestrial planets based on its circumference and diameter as well. At the equator, Earth's circumference is 24,901.55 miles (40,075.16 km). It is slightly smaller between the North and South poles at 24,859.82 miles (40,008 km). Earth's diameter at the poles is 7899.80 miles (12,713.5 km), while it is 7926.28 miles (12,756.1 km) at the equator. For comparison, the largest planet in Earth's solar system, Jupiter, has a diameter of 88,846 miles (142,984 km).

3. *Structure of the Earth.*

At the earth's center is an inner core, and its radius is 1070 km. It is solid and very dense because it exists under extremely high pressure; it is also magnetized, rich in iron and nickel, and very hot (5000 °C). The inner core is surrounded by a transition zone about 700 km thick, which is in turn surrounded by a 1700 km thick layer of liquid material, and together they form outer core. The liquid material of the outer core is similar in composition to the inner core but cooler (4000 °C). The next largest mass of material of any of the layers is about 70% of the earth's volume. This layer is 2835 km thick; it is less dense than the core, still cooler (1500–3000 °C), and composed of magnesium-iron silicates.

The outer part of the mantle is thought to be rigid, while the inner region is deformable and flows slowly over the deeper mantle. The earth's outermost layer is cold, rigid, and thin (10–65 km), which is the crust. The crust is divided into SIAL and SIMA layers. The SIAL is the upper layer that contains mainly silicon and aluminum elements. SIMA is the lower layer that contains silicon and magnesium elements. SIMA layer forms the bottom of the ocean and is known as pyro sphere.

The boundary between the crust and the mantle is the Mohorovicic discontinuity named for its discover (Mohorovicic) and usually called the Moho. The Moho is a boundary at which there is sudden change in the chemical composition and the speed of seismic waves. The mantle just below the crust is rigid, solidified, basalt-type rock, fused to the crust but at the same time separated from it by the Moho. This rigid layer of crust and upper mantle is the lithosphere. The subregion of the mantle extending about 250 km below the lithosphere is the asthenosphere. The lithosphere is less dense than the asthenosphere, and both continental and oceanic lithosphere float on the asthenosphere.

There is a fundamental difference between the crust under the land and under the ocean both in thickness and composition. The continental crust averages about 35 km in thickness, and the ocean crust averages about 11 km including the overlying water. The main rock type of the continents is granite, while that of the oceanic crust is gabbro. Granites and gabbro are both rocks that have formed from the cooling of magma. Continental crust has a density of 2.8 g/cm^3, and oceanic crust has a density of 3.0 g/cm^3.

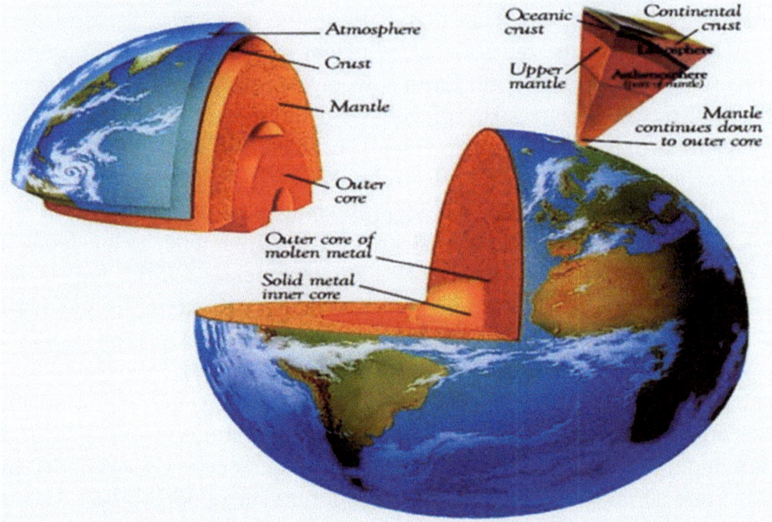

8.1.7   Concept of Latitude and Longitude

Latitude and longitude are geographical coordinates used to specify locations on the Earth's surface. These coordinates are essential for navigation, cartography, and various location-based applications.

Latitude is the angular distance measured north or south from the equator, which is the imaginary line that circles the Earth horizontally midway between the North and South Poles. Latitude is measured in degrees, minutes, and seconds and ranges from 0° at the equator to 90°N (North Pole) and 90°S (South Pole). Latitude influences climate, as regions near the equator receive more direct sunlight, resulting in a warmer climate, while polar regions experience colder temperatures.

Longitude is the angular distance measured east or west from the Prime Meridian, which is an imaginary line running from the North Pole to the South Pole through Greenwich, London. Longitude is measured in degrees, minutes, and seconds and ranges from 0° to 180°E (eastward) and 0° to 180°W (westward) from the Prime Meridian. Longitude is crucial for determining time zones. Every 15 degrees of longitude corresponds roughly to 1 h of time difference.

8.1.8   Cartography

Cartography is the art and science of creating maps. It involves the design, production, and interpretation of maps, which are graphical representations of spatial information. Cartographers use various techniques to convey geographical information effectively. Cartography is a dynamic field that has evolved with advancements in technology and changes in society's mapping needs. Cartographers play a vital role in creating maps that inform decision-making, aid navigation, and convey spatial information to a diverse audience. The key aspects of cartography involve the following:

1. Map Design.

   Cartographers must choose an appropriate scale for the map to ensure that features are represented accurately in relation to their real-world size. Cartographers use symbols, colors, and patterns to represent different features on a map. This includes natural features, human-made structures, and cultural or thematic information. Proper labeling of geographic features, including place names, rivers, and boundaries, is crucial for map interpretation.

2. Map Projections.

   Maps represent a three-dimensional Earth on a two-dimensional surface, which introduces distortions. Cartographers use various map projection methods to minimize distortion in specific areas, shapes, or distances. Examples of map projections include the Mercator, Robinson, and Peters projections, each with its own strengths and weaknesses.

3. Map Elements.

   Maps typically include a title that conveys the map's purpose and a legend that explains the symbols and colors used on the map.

   North Arrow: Indicates the orientation of the map, usually pointing toward the geographic north.
   Scale Bar: Provides a visual representation of distances on the map.

4. Map Types.

   Topographic Maps: Detail the physical features of an area, including contours, elevation, and natural and human-made features.
   Thematic Maps: Emphasize a particular theme or topic, such as population distribution, climate zones, or land use.
   Cadastral Maps: Show land ownership and property boundaries.

   Nautical and Aeronautical Charts: Designed for navigation, showing water depths, coastlines, and navigation aids.

8.1.9 Projections and Landscape

Map projections and the landscape are crucial for accurate representation and interpretation of Earth's surface features. Map projections are methods used to represent a three-dimensional Earth on a two-dimensional map, while landscape characteristics influence the way remote sensing data is acquired and analyzed. The choice of map projections and consideration of landscape characteristics are vital in remote sensing applications. They impact how data is collected, processed, and represented, influencing the accuracy and effectiveness of analyses in fields such as environmental monitoring, agriculture, and urban planning.

8.1.10 Map Projections in Remote Sensing

Map projections are methods used to represent the curved surface of the Earth on a flat, two-dimensional map. Because the Earth is a three-dimensional object, any flat representation introduces distortions in terms of distance, area, shape, and direction. Different map projections are designed to minimize specific types of distortion based on the purpose of the map and the region it represents. Remote sensing data, often collected in a cylindrical or satellite perspective, may require specific projections for accurate representation on maps. Projections are chosen based on the study area's size, shape, and purpose, as well as the type of remote sensing data being used. The most common types of map projections are as follows:

Mercator projection: Preserves angles and shapes but distorts size, particularly at high latitudes. It is often used for navigation particularly for marine charts, because lines of constant compass bearing are straight lines on the map.

Lambert Conformal Conic Projection: Minimizes distortion in shape and area within a specified region, making it suitable for mapping mid-latitude areas such as large countries or continents.

Universal Transverse Mercator (UTM): Divides the Earth into zones, each with its own projection. It minimizes distortion within each zone and is often employed in mapping large areas with an east-west extent.

8.1.11   Landscape Characteristics in Remote Sensing

The landscape has a significant impact on remote sensing, influencing the acquisition, interpretation, and analysis of data collected by satellite or airborne sensors. Understanding the characteristics of the landscape is crucial for obtaining accurate and meaningful information. The scale of remote sensing data influences the level of detail captured. Understanding the scale is essential for accurate interpretation and analysis of landscape features. Some of the landscape factors affecting the data of remote sensing are as follows:

1. *Terrain and Topography.*

   The landscape's elevation, slope, and aspect impact the way remote sensing data is collected and interpreted. Topographic maps are often used in conjunction with remote sensing data to understand terrain features.

2. *Land Cover and Land Use.*

   Different landscapes exhibit distinct land cover and land use patterns. Remote sensing data is used to classify and map these patterns, contributing to applications like agriculture monitoring, urban planning, and environmental assessment.

3. *Vegetation and Canopy Structure.*

   Vegetation cover and canopy structure influence the reflectance properties observed in remote sensing data. Different vegetation types have unique spectral signatures, allowing for identification and monitoring.

4. *Water Bodies.*

   Remote sensing is used to map and monitor water bodies, including rivers, lakes, and oceans. Variations in water characteristics such as turbidity and temperature can be observed from satellite or airborne sensors.

5. *Urban Areas.*

   Urban landscapes pose unique challenges in remote sensing due to the mix of impervious surfaces, buildings, and vegetation. High-resolution imagery and advanced classification methods are often employed to map urban features.

6. *Natural Hazards and Disasters.*
   Remote sensing plays a crucial role in assessing and monitoring natural hazards such as earthquakes, floods, and wildfires. It provides timely information for disaster response and management.
7. *Temporal Changes.*
   Landscape dynamics, including seasonal changes and human-induced alterations, can be monitored over time using remote sensing time-series data. Change detection analysis helps identify land cover changes.

8.2   Geographic Information (GI)

It refers to data that is inherently spatial and can be related to specific locations on the Earth's surface. It involves the collection, storage, processing, analysis, and visualization of spatial data to derive meaningful information about the Earth's features and phenomena. Geographic information (GI) refers to information about the Earth's surface that is represented by spatial data, often in the form of maps, images, and digital datasets. It includes both natural and human-made features and their attributes.

GI focuses on the spatial relationships between geographic features. It involves capturing and organizing information in a way that relates to specific locations on the Earth's surface. In addition to spatial information, GI incorporates attributes or characteristics associated with geographic features. For example, a geographic information system (GIS) may store not only the location of a city but also its population, elevation, and climate data.

Geographic information plays a vital role in various disciplines, providing a spatial context for decision-making and analysis. The integration of spatial data with advanced tools and techniques enhances our understanding of the Earth's complex systems and supports informed decision-making processes.

8.2.1   Tools of GI

Geographic Information System (GIS): A system designed to capture, store, analyze, and visualize spatial data. GIS integrates hardware, software, data, and people to support decision-making processes.

Global Positioning System (GPS): A satellite-based navigation system that enables accurate location determination on the Earth's surface.

Remote Sensing: The collection of information about the Earth's surface using sensors on aircraft or satellites. Remote sensing provides imagery and data for GIS applications.

Web Mapping Tools: Platforms and applications that enable the creation and sharing of interactive maps on the internet.

8.2.2   Techniques

Spatial Analysis: GIS allows for the analysis of spatial relationships, patterns, and trends. Techniques include overlay analysis, buffer analysis, spatial interpolation, and network analysis.

Data Modeling: Creating models to represent the real-world processes and relationships within a GIS. This helps in simulating scenarios and predicting outcomes.

Data Capture: The process of collecting new geographic data or updating existing datasets. Techniques include GPS surveys, digitization from maps, and remote sensing.

Cartography: The art and science of mapmaking. Cartographic techniques are used to design and produce maps for visualization and communication purposes.

8.2.3   Applications

Urban Planning: GIS is used for land-use planning, infrastructure development, and analyzing the impact of urban growth.

Environmental Management: GIS aids in monitoring and managing natural resources, tracking environmental changes, and supporting conservation efforts.

Emergency Management: GIS helps in disaster response, risk assessment, and planning for emergency situations.

Healthcare: Spatial analysis is applied to disease mapping, healthcare facility location planning, and analyzing the spread of diseases.

Transportation: GIS supports route planning, traffic management, and the analysis of transportation networks.

8.2.4   Precision Farming

Geographic information (GI) plays a significant role in precision farming, also known as precision agriculture. Precision farming leverages advanced technologies, including GIS, to optimize agricultural practices, increase efficiency, and enhance overall crop management. Precision farming, enabled by GI, allows farmers to adopt a more site-specific and data-driven approach to agriculture. By leveraging spatial information, farmers can optimize resource use, enhance productivity, and promote sustainability in agricultural practices.

Here are several ways in which GI is utilized in precision farming:

1. Spatial Mapping.

   Soil Variability Mapping: GIS is used to create detailed maps of soil properties and characteristics, including nutrient levels, moisture content, and texture. This information helps farmers understand spatial variability within a field.

Field Boundaries and Land Parcel Mapping: Accurate mapping of field boundaries and land parcels allows farmers to organize and manage their agricultural land effectively.

2. Variable Rate Technology (VRT).

Variable Rate Application (VRA): GIS is employed to analyze spatial data and determine optimal rates for inputs such as fertilizers, pesticides, and irrigation. VRA allows for the precise application of inputs based on the specific needs of different areas within a field.

3. Crop Monitoring.

Satellite Imagery and Remote Sensing: GIS is integrated with satellite imagery to monitor crop health, detect diseases, and assess overall vegetation conditions. Remote sensing data helps identify areas of stress, enabling timely intervention.

4. Precision Planting.

Planting Density Optimization: GIS assists in determining optimal planting densities across a field by considering soil characteristics and historical performance data. This ensures that seeds are planted at the right spacing for optimal growth.

5. Yield Monitoring.

Harvest Yield Mapping: GIS is used to create yield maps by recording and analyzing data from yield monitors during harvest. This helps farmers identify areas of the field with high or low yields, contributing to future decision-making.

6. Weather and Climate Data Integration.

Weather Stations and Sensors: GIS is utilized to integrate weather and climate data with spatial information. This integration helps farmers make informed decisions related to irrigation scheduling, pest management, and other weather-dependent activities.

7. Farm Management Software.

GIS-Based Decision Support Systems: Farm management software often incorporates GIS to provide decision support tools. These tools help farmers analyze spatial data, plan activities, and make informed decisions about crop management.

8. Precision Livestock Farming.

Livestock Tracking and Management: GIS is applied in precision livestock farming to track the movement and behavior of animals. It helps optimize grazing patterns, manage feeding, and monitor the health of livestock.

9. Resource Optimization.

   Water Management: GIS aids in optimizing irrigation practices by analyzing soil moisture levels and topographic features. This helps reduce water usage and ensures efficient water distribution.
   Nutrient Management: GIS assists in precise nutrient application by considering soil variability, reducing the risk of overapplication and minimizing environmental impact.

10. Record Keeping and Reporting.

   Historical Data Analysis: GIS is used to maintain historical records and analyze trends over time. This information is valuable for long-term planning and decision-making.

8.2.5 Yield Monitoring

GI, particularly through the use of GIS, plays a crucial role in yield monitoring in agriculture. Yield monitoring involves the collection and analysis of data related to crop productivity, and GIS enhances this process by providing spatial context to the information. Here are several ways in which GI is utilized in yield monitoring:

1. Yield Mapping.

   Spatial Representation: GIS is used to create detailed yield maps that represent variations in crop productivity across a field. These maps visually display the spatial distribution of yields, allowing farmers to identify high-yielding and low-yielding areas.

2. Integration with Precision Agriculture.

   Variable Rate Technology (VRT): GIS is integrated with VRT to optimize input application rates based on the variability observed in yield maps. This ensures that inputs like fertilizers and pesticides are applied at variable rates to match the specific needs of different areas within a field.

3. Data Collection.

   Combine Harvesters and GPS Technology: GPS-equipped combine harvesters collect real-time yield data as crops are harvested. This data is spatially referenced, allowing for accurate mapping of yield variations.

4. Spatial Analysis.

   Spatial Patterns and Trends: GIS enables farmers to conduct spatial analysis on yield maps to identify patterns and trends. This information can be used to understand the factors influencing yield variability, such as soil characteristics or drainage patterns.

5. Decision Support.

> Site-Specific Decision-Making: GIS provides a platform for integrating yield data with other spatial layers, such as soil maps, topography, and historical information. Farmers can make informed, site-specific decisions for future planting, irrigation, and nutrient management based on this integrated data.

8.2.6 Fertilizer Recommendation

Fertilizer recommendation through remote sensing involves using satellite or aerial imagery, along with other spatial data, to assess crop health, detect nutrient deficiencies, and optimize fertilizer applications. Analyzing satellite imagery over time allows for monitoring changes in crop health throughout the growing season by detecting anomalies or trends, which help make timely fertilizer recommendations. The varied reflectance patterns in different spectral bands help identify nutrient stress or deficiencies in crops. Unhealthy plants may exhibit distinct spectral signatures compared to healthy ones. Variable rate technology (VRT) involves creating prescription maps based on remote sensing data. These maps guide variable-rate fertilizer application, adjusting rates based on spatial variations in crop health. Integrating remote sensing data with soil data helps correlate crop health with soil properties. Identifying relationships assists in refining fertilizer recommendations. Remote sensing supports precision agriculture by enabling site-specific fertilizer recommendations. The goal is to optimize nutrient inputs based on spatial variability within a field. Fertilizer recommendation through remote sensing is a dynamic process that requires a combination of technological understanding, agronomic knowledge, and practical implementation skills. It involves leveraging advanced technologies to enhance precision and efficiency in nutrient management, ultimately contributing to improved crop productivity and sustainability.

8.2.7 Global Information System (GIS)

Geographic Information System (GIS) plays a crucial role in agriculture by integrating spatial data and providing tools for analysis, visualization, and decision-making. The integration of GIS with technologies like remote sensing, GPS, and IoT enhances its capabilities in agriculture, providing farmers with valuable information for sustainable and efficient farming practices.

> Mapping and Monitoring: GIS is used to create detailed maps of fields, showing variations in soil types, moisture levels, and crop health. These maps help farmers make informed decisions on irrigation, fertilization, and pest control.

Variable Rate Technology (VRT): By integrating GIS with VRT, farmers can apply inputs (such as water, fertilizers, and pesticides) at variable rates across a field based on spatial variations, optimizing resource use and improving yields.

Crop Management: GIS helps in analyzing soil types, topography, and climate data to determine the most suitable crops for specific areas. GIS aids in planning crop rotation strategies by considering factors such as nutrient depletion, pest cycles, and soil health.

Water Management: GIS is used to optimize irrigation by assessing soil moisture levels, topography, and weather patterns. This helps in efficient water use and prevents overirrigation. GIS is employed in managing watersheds to monitor water quality, assess runoff, and plan conservation practices.

Disease and Pest Management: GIS helps in monitoring and analyzing the spatial distribution of pests and diseases. Early detection allows for targeted interventions, reducing the need for widespread pesticide application. GIS can assess the risk of pest and disease outbreaks based on environmental factors, helping farmers implement preventive measures.

Disaster Management: GIS is instrumental in assessing and managing the impact of natural disasters on agriculture, helping authorities and farmers respond effectively to events like floods, droughts, or wildfires.

8.2.8   *Global Positioning System (GPS)*

Global Positioning System (GPS) technology has become an integral part of modern agriculture, providing farmers with accurate location information and enabling precision farming practices.

Precision Agriculture: GPS is used to create accurate field maps, allowing farmers to understand the variability in their fields in terms of soil types, moisture levels, and crop health. By integrating GPS with VRT, farmers can apply inputs such as fertilizers, pesticides, and water at variable rates across a field, optimizing resource use and improving crop yields.

Navigation and Guidance: GPS assists farmers in precisely navigating their tractors and other machinery, ensuring optimal spacing between rows and reducing overlaps during planting, spraying, and harvesting operations. GPS-enabled automated steering systems help farmers maintain straight and consistent paths while operating tractors and other equipment, improving efficiency and reducing operator fatigue.

Mapping and Monitoring: GPS, when combined with other sensors and technologies, allows farmers to monitor crop health, growth stages, and yield variations across the field. GPS is used to create yield maps, providing insights into the variations in crop yield within a field. This information helps in identifying patterns and optimizing future planting and management practices.

Irrigation Management: GPS-guided systems help in steering and controlling irrigation equipment, ensuring precise water application and reducing water wastage. GPS is used to map variations in soil moisture levels, allowing farmers to make informed decisions on irrigation scheduling.

Asset Management: GPS is used to track the location of farm equipment, allowing farmers to optimize equipment usage, reduce downtime, and enhance overall fleet management. GPS-enabled collars or tags help in tracking the movement of livestock, preventing loss, and optimizing grazing patterns.

Precision Pest Control: GPS data assists in the precise application of pesticides, targeting specific areas where pest pressures are higher, reducing the overall use of chemicals.

The integration of GPS technology with other data sources, such as GIS, remote sensing, and sensor networks, enhances its capabilities and contributes to the development of smart and sustainable agriculture practices.

8.2.9 *Crop Simulation Models*

Crop simulation models, also known as crop models or agricultural models, are computational tools that simulate the growth and development of crops under various environmental conditions. These models use mathematical equations and algorithms to represent the complex interactions between the crop, soil, climate, and management practices. Crop simulation models are valuable tools for researchers, agronomists, and farmers as they can provide insights into crop performance, help optimize agricultural practices, and contribute to decision-making processes.

"Crop simulation models are quantitative tools that predict the growth, development, and yield of crops as a function of the environment, soil, and management practices" (Ritchie and Otter-Nacke (1985). According to Hammer et al. (2010), crop simulation models are mathematical representations of plant growth and development that simulate the effects of environment, soil, and management on crop production.

The central role of crop simulation models is predicting and understanding the complex interactions between crops and their environment, including soil conditions and management practices. These models are valuable for decision support in agriculture, enabling farmers and researchers to explore various scenarios and optimize farming practices for improved productivity and sustainability.

8.2.10 *Nanotechnology*

Nanotechnology involves manipulating materials at the nanoscale, typically with dimensions less than 100 nm. In agriculture, nanotechnology offers innovative solutions to enhance various aspects of crop production, soil management, and pest control. Nanotechnology in agriculture holds promise for addressing various challenges, and it is essential to consider potential environmental and safety concerns associated with the use of nanomaterials. Various applications of nanotechnology in agriculture are as follows:

Nanofertilizers: Nanofertilizers are nano-sized particles containing essential nutrients for plants. It helps in enhanced nutrient absorption, controlled release of nutrients, reduced environmental impact, and increased nutrient use efficiency.

Nanosensors: Nanosensors are devices at the nanoscale that can detect and measure specific parameters like real-time monitoring of soil conditions, water quality, and plant health, facilitating precise resource management.

Nanopesticides: Nanopesticides are nanoformulations of pesticides designed to improve delivery and efficacy. Increased pesticide stability, reduced environmental impact, targeted delivery to pests, and lower application volumes.

Nanomaterials for Soil Remediation: Nanomaterials, such as nanoparticles, can be used to remediate contaminated soils leading to removal of pollutants, improved soil structure, and increased efficiency in soil remediation processes.

Nanoscale Sensors for Plant Health Monitoring: Small-scale sensors at the nanoscale can monitor plant health parameters like early detection of plant stress, diseases, or nutrient deficiencies, enabling timely interventions.

Improved Seed Coatings: Nanotechnology is applied to create enhanced seed coatings, and it helps to improve seed germination, enhanced protection against pests and diseases, and increased seedling vigor.

8.3 Agri-Informatics

Agricultural informatics, often referred to as agri-informatics or agroinformatics, is a multidisciplinary field that involves the use of information and communication technologies (ICT) to enhance efficiency, productivity, and sustainability in agriculture. It encompasses the application of various technologies, data management systems, and computational tools to collect, process, analyze, and disseminate information related to agricultural practices. The goal of agricultural informatics is to make informed and data-driven decisions for optimizing farming operations. Agricultural informatics plays a crucial role in modernizing agriculture by leveraging technology to address challenges, enhance productivity, and promote sustainable practices. It enables farmers, researchers, and other stakeholders to make informed decisions based on data-driven insights.

Data Collection and Sensing: The uses of satellite or aerial imagery to monitor crop health, assess soil conditions, and identify pest or disease outbreaks. It is done by utilizing on-field sensor technologies to collect real-time data on parameters like soil moisture, temperature, and nutrient levels.

Precision Agriculture: Integration of GPS for accurate mapping, navigation, and precise application of inputs, such as fertilizers and pesticides utilizing VRT by adjusting input application rates based on spatial variability detected by sensors and GPS, optimizing resource use.

Farm Management Information Systems (FMIS): Utilizing databases and information systems to store, organize, and manage data related to crops, soils, weather, and farm operations.

Agro-Meteorological Information Systems: Integrating weather data to provide accurate and timely forecasts, helping farmers plan activities and mitigate risks associated with extreme weather events.

Internet of Things (IoT): Deploying IoT devices for real-time monitoring and control of agricultural processes, improving efficiency and resource use.

Drones and Unmanned Aerial Vehicles (UAVs): Using drones for aerial surveys and mapping of fields, helping farmers identify issues such as crop diseases, nutrient deficiencies, and pest infestations. UAVs can also be employed to capture high-resolution images for detailed crop monitoring.

8.3.1 Use of ICT in Agriculture

Information and Communication Technology (ICT) plays a vital role in modernizing agriculture by introducing advanced tools and systems to improve efficiency, productivity, and sustainability. Remote sensing satellites provide high-resolution images for monitoring crop health, identifying pest infestations, and assessing overall field conditions.

- Precision agriculture utilizing GPS is used for accurate mapping of fields, enabling precision farming practices such as VRT for optimized input application.
- Mobile applications and online platforms can be utilized to provide real-time weather information, forecasts, and alerts to help farmers plan activities and mitigate risks. Farming practices integrated with climate data make the agricultural operation more sustainable and resilient, which is also termed as climate smart agriculture.
- Apps that provide farmers with market prices, trends, and information help them to make decisions about selling their produce.
- Extension Services: Mobile apps and messaging services for farmers to access agricultural extension services, advice, and training.
- It helps the farmers to monitor soil conditions, weather, and crop health in real time by utilizing IoT sensors and hence becomes a component of smart farming.
- Drones equipped with cameras and sensors are deployed for monitoring crops, detecting diseases, and assessing field conditions.
- Using ICT to establish early warning systems for pest and disease outbreaks, enabling timely interventions.
- Government portals and e-agriculture initiatives for streamlined access to agricultural services, subsidies, and information.

- Facilitating online transactions and payments for agricultural inputs, produce, and financial transactions.
- Using ICT to establish transparent and traceable supply chains, ensuring food safety and quality.
- Blockchain Technology: Utilizing blockchain for secure and transparent recording of transactions, ensuring the authenticity and origin of agricultural products.

8.3.2 Agri Input Management through Computer-Controlled Devices

Agricultural input management through computer-controlled devices, often associated with precision agriculture, involves the use of technology to optimize the application of inputs such as fertilizers, pesticides, water, and seeds. These computer-controlled devices leverage various technologies, including GPS, sensors, actuators, and data analytics, to enhance the efficiency, accuracy, and sustainability of agricultural practices. Here are some examples of how computer-controlled devices are used in agri-input management:

Variable Rate Technology (VRT): VRT involves adjusting the rate of input application (e.g., fertilizers, pesticides) based on the variability of field conditions leading to optimized resource use, reduced input wastage, and improved crop yields. GPS-enabled devices and sensors analyze field data (soil fertility, moisture levels, etc.) in real time, allowing for precise application of inputs at variable rates across the field.

Automated Planting Systems: Computer-controlled planters use GPS and sensors to optimize seed placement and spacing during planting hence improved seed germination, uniform crop stand, and efficient use of seeds. GPS guidance systems ensure accurate row spacing and seed placement, optimizing planting patterns.

Precision Irrigation Systems: Precision irrigation involves delivering water to crops based on their specific needs, considering factors like soil moisture and crop type leading to water-use efficiency, reduced water wastage, and improved crop performance. Sensors in the field measure soil moisture levels and computer-controlled irrigation systems adjust water application rates accordingly.

Automated Fertilizer Application: Computer-controlled fertilizer applicators adjust the application rate of fertilizers based on real-time field conditions and hence enhanced nutrient management, reduced environmental impact, and improved crop nutrition. Sensors and GPS technology analyze soil nutrient levels and variability, optimizing the application of fertilizers.

Smart Sprayers for Pest Control: Computer-controlled sprayers adjust the application of pesticides based on crop conditions and pest pressures with aim of targeted pesticide application, reduced chemical use, and minimized impact on nontarget areas. Sensors and cameras identify areas with higher pest densities, and the sprayer adjusts the spray pattern accordingly.

Automated Harvesting Systems: Computer-controlled harvesting equipment optimizes the harvest process, ensuring efficiency, minimizing losses and improved overall yield. GPS and sensors guide harvesting equipment, optimizing the harvesting route, and adjusting machinery settings.

The integration of computer-controlled devices in agriculture not only optimizes input management but also contributes to sustainable farming practices by minimizing resource wastage and environmental impact. These technologies enable farmers to make data-driven decisions, enhance efficiency, and improve the overall productivity of their operations.

8.3.3   Use of Computer Models for Understanding Plant Processes

Computer models for understanding plant processes are an excellent way to engage them in hands-on, experiential learning. Computer models provide a dynamic and interactive platform for students to explore complex biological concepts, simulate plant processes, and gain a deeper understanding of the interactions within plant systems.

1. *Simulation of Photosynthesis.*

   By using software or online tools that simulate photosynthesis processes and to explore factors influencing photosynthetic rates, such as light intensity, temperature, and carbon dioxide concentration. The impact of environmental conditions on photosynthesis and relate it to real-world scenarios, such as the effects of climate change or variations in growing conditions.

2. *Crop Growth Models.*

   Crop growth simulation models that depict the growth stages of specific crops. The impact of factors like nutrient availability, water stress, and pest infestation on crop development. The experiment with different management practices, such as varying planting dates or adjusting irrigation schedules, and observe the outcomes on crop growth.

3. *Ecological Modeling.*

   The ecological models that simulate interactions between plants and their environment, including competition for resources, predator-prey relationships, and symbiotic interactions. The ecological implications of different plant strategies and how these models contribute to our understanding of ecosystems.

4. *Dynamic Plant Physiology Simulations.*

   Utilization of dynamic simulation tools that allow to manipulate variables related to plant physiology, such as transpiration rates, nutrient uptake, and hormone regulation. The link simulations to laboratory experiments helps to connect theoretical knowledge with practical observations.

5. *Online Plant Physiology Simulators.*
  The exploration of freely available online simulators that allow students to experiment with plant processes. These simulators often come with user-friendly interfaces and interactive features. The limitations and assumptions of the models, fostering critical thinking and a deeper understanding of the scientific modeling process.

8.3.4 Apps for Agro-Advisories

Governments in various countries often develop and promote agriculture-related apps to provide farmers with agro-advisories, weather forecasts, market prices, and other relevant information. Here are some examples of government-backed apps for agro-advisories:

M-Kisan (India): M-Kisan offers personalized agro-advisories, information on government schemes, market prices, and weather updates. It is available in multiple regional languages.

FASAL (India): FASAL provides location-specific weather forecasts, crop health monitoring, and agro-advisories to farmers.

Agricultural Meteorology Division (AMD) Agro-Met Advisory Bulletins (India): Provides agro-meteorological advisories including weather forecasts, pest and disease warnings, and crop-specific recommendations.

AgriMarket (India): AgriMarket provides market prices, demand forecasts, and related information to help farmers make informed decisions.

Krishi Bhagya (Karnataka, India): Krishi Bhagya offers weather forecasts, crop-specific agro-advisories, and information on government schemes.

Krishi Gyan Sagar (Uttar Pradesh, India): Krishi Gyan Sagar provides agro-advisories, information on agricultural practices, and weather updates in Hindi.

Farm Advice (Australia): Provides agro-advisories, information on farm management, and resources for Australian farmers.

FarmSmart (Canada): Offers agro-advisories, crop management information, and resources for Canadian farmers.

AgriBus-NAVI (Japan): Provides agricultural machinery guidance, GPS-based field mapping, and other agro-advisory services.

eKisaan (Pakistan): eKisaan offers agro-advisories, weather updates, and information on crops and pests for Pakistani farmers.

These government-backed agro-advisory apps aim to empower farmers with timely and relevant information, ultimately contributing to improved agricultural practices and better yields. Users should verify the authenticity of these apps by downloading them from official government sources or authorized app stores.

8.3.5   Crop Weather Calendars

Crop weather calendars are tools that help farmers and agricultural practitioners plan and manage their farming activities based on local weather conditions and the specific needs of different crops. These calendars provide a visual guide that aligns key agricultural practices with the seasons and climatic conditions, helping optimize resource use and improve overall crop management. Here are the typical components found in crop weather calendars:

1. *Planting Dates.*
   Indicate the optimal time to sow or transplant crops based on historical weather patterns. Considerations include soil temperature, frost dates, and the preferred growing conditions of specific crops.
2. *Growing Periods.*
   Highlight the typical duration of the growing season for different crops. This information helps farmers plan for the stages of crop development, from germination to maturity.
3. *Watering Schedule.*
   Provide guidelines for irrigation based on expected rainfall patterns and the water needs of different crops at various growth stages. This helps optimize water use efficiency.
4. *Fertilization Timing.*
   Outline the recommended times for applying fertilizers, taking into account nutrient requirements during critical growth phases. Factors such as soil nutrient levels and weather conditions are considered.
5. *Pest and Disease Management.*
   Include information on the emergence and activity of pests and diseases. This helps farmers plan for preventive measures, such as applying pesticides or adopting integrated pest management practices.
6. *Harvest Dates.*
   Specify the expected time for harvesting each crop. This information is crucial for coordinating labor, equipment, and storage facilities, ensuring efficient harvest operations.
7. *Climatic Risk Management.*
   Highlight periods of increased climatic risk, such as the likelihood of extreme weather events (e.g., storms, frosts, or heatwaves). Farmers can take precautions to minimize potential damage to crops.
8. *Crop Rotation Planning.*
   Suggest ideal crop rotation sequences to improve soil health, manage pests, and optimize nutrient cycling. Crop rotations are designed to complement weather patterns and reduce the risk of disease buildup.

9. *Cover Crop Planting Dates.*

   If cover crops are part of the farming strategy, the calendar may include recommended dates for planting cover crops to improve soil structure, control erosion, and enhance nutrient cycling.

10. *Critical Growth Periods.*

   Identify key growth phases for each crop, such as flowering, fruiting, and grain filling. Farmers can tailor their management practices to support crops during these critical periods.

Crop weather calendars are often developed at the local or regional level, taking into account the specific climate and soil conditions of the area. They are valuable tools for promoting sustainable and climate-smart agriculture by aligning farming activities with the natural cycles and variations in weather patterns. Additionally, they help farmers adapt to changing climates and mitigate risks associated with extreme weather events.

MCQs

1. Remote sensing is.

   (a) Acquiring physical data through direct contact.
   (b) Acquiring information about objects or areas without direct contact.
   (c) Processing pictures and other electromagnetic records.
   (d) None of the above.

2. In the context of remote sensing, what is GIS?

   (a) Global Information System.
   (b) Geographic Information System.
   (c) General Imaging System.
   (d) Gravitational Information System.

3. The two broad types of sensing systems.

   (a) Short-range and long-range sensing systems.
   (b) Active sensing system and passive sensing system.
   (c) Microwave and infrared sensing systems.
   (d) Solar and terrestrial sensing systems.

4. Cartography means.

   (a) The study of geographical features.
   (b) The design, production, and interpretation of maps.
   (c) The analysis of spatial data.
   (d) The exploration of navigation techniques.

5. The characteristic feature of the Mercator projection.

    (a) Minimizes distortion in size at high latitudes.
    (b) Preserves size but distorts shapes, especially at high latitudes.
    (c) Minimizes distortion in both size and shape.
    (d) Used primarily for mid-latitude areas.

6. UTM stand for in the context of map projections.

    (a) Universal Time Meridian.
    (b) Unifying Terrain Mapping.
    (c) Universal Transverse Mercator.
    (d) Unified Topographic Model.

7. VRT stand for in the context of precision agriculture.

    (a) Variable Resource Technology.
    (b) Variable Rate Technology.
    (c) Visual Remote Targeting.
    (d) Virtual Reality Too.

8. The central role of crop simulation models.

    (a) Predicting financial market trends.
    (b) Simulating environmental conditions.
    (c) Understanding interactions between crops and their environment.
    (d) Modeling atmospheric phenomena.

9. The dimensions typically associated with nanoscale materials.

    (a) More than 100 nm.
    (b) Exactly 100 nm.
    (c) Less than 100 nm.
    (d) Exactly 1 nm.

10. The integration of agricultural operations with climate data, making them more sustainable and resilient is termed as

    (a) Climate-sensitive farming.
    (b) Weather-dependent agriculture.
    (c) Climate-conscious farming.
    (d) Climate-smart agriculture.

References

Hammer, G. L., van Oosterom, E., McLean, G., Chapman, S. C., Zheng, B., Doherty, A., & Chapman, R. (2010). Adapting APSIM to model the physiology and genetics of complex adaptive traits in field crops. *Journal of Experimental Botany, 61*(8), 2185–2202.

Lintz, J., & Simonett, D. S. (1976). Sensors for spacecraft. In *Remote sensing of environment* (pp. 323–343). Addison-Wesley.

Mather, P. M., & Koch, M. (2011). *Computer processing of remotely-sensed images: An introduction.* Wiley.

Ritchie, J. T., & Otter-Nacke, S. (1985). *A user-oriented model of the soil-plant-atmosphere system—Rice simulation model (RiceSim)* (Vol. 1, pp. 79–89). International Rice Research Institute (IRRI).

Chapter 9
Protected Farming

Abstract Protected farming refers to the modification of the natural environment of the crop to achieve optimum plant growth. This technology is gaining popularity day by day and is used for vegetables, flowers, and other high value perishable crops. The most practical method of achieving objectives of protected farming is greenhouse. This chapter presents details about the history, origin classification, and designs of greenhouse. Later on, chapter moves on to heating and cooling under the greenhouse conditions. Separate sections are dedicated to cultivation procedure of crops, fertilization, and irrigation in greenhouse.

Keywords Greenhouse · Mulching · Evaporative cooling · Sterilization · Quonset · Fertigation · Irrigation

9.1 What Is Protected Farming?

The process of cultivating crops in a controlled environment is known as protected farming. As a result, it is possible to control the temperature, humidity, light, and other variables according to the needs of the crop (Keshwania et al., 2021). All of the necessary farming conditions can be easily controlled as needed. The farming method is healthier and produces a large amount of food under uniform conditions. There are many different types of protected agriculture, including forced ventilated greenhouses, naturally ventilated polyhouses, insect-proof net houses, shade net houses, plastic tunnels and mulching, raised beds, trellising, and drip irrigation.

9.2 Objectives of Protected Farming

The main objectives of protected farming are as follows:

- The primary goal of the farming technique is to shield the plants from abiotic stress, which includes both physical and nonliving factors like temperature, water availability (or lack thereof), hot and cold waves, and biotic factors like the occurrence of pests and diseases, among other things. All of the natural and climatic conditions can be easily adjusted for plants thanks to protected cultivation in India.
- This technique uses little water, while weed growth is kept to a minimum. When growing the plant, this method conserves a lot of water.
- The production per area increases under protected cultivation, generating high output and additional income.
- This method reduces the use of pesticides in crop production while maintaining crop quality.
- Protected farming propagates healthy, uniform, and disease-free planting material, which improves germination percentages and provides better hardening.
- This farming method produces disease-free yield and transplants that are genetically superior.

9.3 Importance of Protected Farming

Agriculture has long been the backbone of India's economy, as we all know. India is self-sufficient in agriculture, ensuring adequate food security and exporting high-quality fruits and vegetables. Nonetheless, the demand for high-quality agricultural production has increased over the last decade, and India has yet to meet it. To meet market demand, Indian agriculture requires new and efficient production technologies that can continually improve farm productivity, profit growth, and respectability (Keshwania et al., 2021).

In order to meet out the food demand in 2050, world production must increase by 70% (FAO, 2020). It can be done by various man-made technologies in the field of agriculture. Protected farming is one of these technologies, and it is now widely used in developed nations like India. This farming technique is a new one that allows for various cropping patterns and climatic variations. India has a significant issue with climatic extremes like floods, droughts, and other climatic abnormalities that regularly result in crop losses or damages, which result in economic losses. So, in order to avoid these all-harmful conditions, the protected cultivation method was developed, which gives Indian farmers better opportunities.

The greenhouse is one of the best examples of commercially used protected cultivation for the production of non-native and off-season vegetables, flowers, and quality seedlings. The greenhouse method significantly increased the economic returns of high-value agricultural products. Any plant can theoretically be grown in a protected environment, but due to economic profitability, management, and production cycles, this method is primarily used to grow cut flowers and vegetables. Because of this, protected horticulture rather than protected agriculture is referred to more frequently (Cedillo & Calzada, 2012).

9.4 Greenhouse

A greenhouse is a structure where plants are grown that has walls and a roof primarily made of transparent material. A greenhouse's interior, which is heated by sunlight, is significantly warmer than the surrounding air, shielding the plants inside from harsh weather. In order to optimize plant growth conditions, modern smart greenhouses connect to a computer and use technology for heating, cooling, lighting, and other functions (Fig. 9.1).

Currently, open fields are used to grow 92% of all plants that are raised by humans. Farmers have had to adapt to the growing conditions mother nature has provided them with, ever since agriculture was first practiced. Man has developed technological means of growing some high value crops by protecting them from the excessive cold and excessive heat in some temperate regions where the climatic conditions are so unfavorable that no crops can be grown. The term for this is greenhouse technology. "Greenhouse technology" is the science of giving plants a healthy environment. In addition, it shields plants from harmful weather elements like wind, cold, precipitation, excessive radiation, extreme heat, insects, and diseases. The plants can be surrounded by a perfect microclimate.

Fig. 9.1 Commercial greenhouse. (Pini et al., 2020)

9.4.1 History and Origin of Greenhouse

Agriculture production within protected structures began in the nineteenth century in France and the Netherlands. This method was used in simple, low-rise glass structures that provided climate protection and were primarily used for the cultivation of ornamental plants. The technology for building greenhouses had advanced by the turn of the twentieth century, primarily as a result of the end of the Second World War. This was especially true in the colder nations of Western Europe, with the Netherlands setting the pace. The structures gradually gained agro-technical systems, aeration options, and accompanying accessories, while the foundations were upgraded to the well-known, conventional heavy steel constructions covered in rigid glass boards. By the end of the 1950s, greenhouse technology had spread to northern and central Europe, extending its influence and advantages to Israel, where a wave of experiments and studies in the area had already started. A brand-new style of structure-covering sheets was made public in the 1960s. The flexible, inexpensive polyethylene sheets were responsible for the conceptual shift in the greenhouse industry. Other types of effective light transition coverings, like polycarbonate (a type of covering made of plastic polymers), simultaneously emerged, leaving the conventional glass covering in their wake.

9.4.2 Classification of Greenhouse

Building a greenhouse is based on investment. The potential for attaining tightly controlled growing conditions increases with the level of technology used. To help people choose the best investment for their needs and budget, the three types of greenhouses below have been defined.

9.4.2.1 Classification on the Basis of Technology Used

- *Low-Technology Greenhouse*: The total height of these greenhouses is less than 3 m. The most prevalent kind is tunnel housing. There are no vertical walls in them. They lack adequate ventilation. This kind of building is simple to erect and reasonably priced. Automation is hardly ever used. The crop potential is still constrained by the growing environment, and crop management is relatively challenging, despite the fact that this type of structure offers some fundamental advantages over field production. Low-level greenhouses typically create an unfavorable environment for growing crops, limiting yields and doing little to lower the incidence of pests and diseases.
- *Medium-Technology Greenhouses*: Vertical walls more than 2 m but less than 4 m tall and a typical total height of less than 5.5 m are characteristics of medium level greenhouses. They may have ventilation in the roof, the side walls, or both.

Medium-level greenhouses typically use varying levels of automation and are covered in either single or double skin plastic film or glass. A reasonable economic and environmental foundation for the industry is provided by medium-level greenhouses, which provide a compromise between cost and productivity. Field production may not always be more efficient than production in medium-level greenhouses. The efficiency of water use is increased by hydroponic systems.

- *High-Technology Greenhouses*: High level greenhouses have walls that are at least 4 m high and have roof peaks that are up to 8 m high. The crop and environmental performance of these structures is superior. High-tech buildings will likely have roof vents and possibly side wall vents as well. Glass, polycarbonate sheeting, single or double plastic film, and others can be used as cladding. Most of the time, environmental controls are automated. The potential for economic and environmental sustainability provided by these structures is enormous. Pesticide usage can be greatly reduced.

9.4.2.2 Classification of Greenhouses Based on Shape

- *Lean-to Type Greenhouse*: When a greenhouse is built up against the side of another structure, a lean-to design is used. It is constructed with one or more of its sides against an existing building. It is frequently affixed to a house, but it could also be affixed to other structures. The building's roof has been extended with the proper greenhouse covering material, and the space is properly enclosed. It typically faces the south. The single- or double-row plant benches with a total width of 7–12 feet are the only options for the lean-to type greenhouse. It may be as long as the building to which it is affixed. For optimal sun exposure, it ought to face that direction (Fig. 9.2).
- *Even-Span Type Greenhouse*: The two roof slopes on an even-span structure have the same pitch and width. It is a full-size standard type of structure. This style of greenhouse is used for smaller greenhouses that are built on flat ground. Two or three rows of plant benches can fit inside. Even-span greenhouses are more expensive than lean-to types, but they can accommodate more plants and have greater design flexibility. The even-span will cost more to heat due to its size and greater amount of exposed glass area. Compared to a lean-to type, the design has a better shape for air circulation to keep temperatures consistent during the winter heating season (Fig. 9.3).
- *Uneven-Span Type Greenhouse*: This type of greenhouse is built on sloping ground. The roofs are of unequal width, making the structure adaptable to hillside side slopes. This type of greenhouse is rarely used nowadays because it cannot be automated. Because the roof is longer on one side, more sunlight can enter the greenhouse without being blocked by side walls (Fig. 9.4).
- *Ridge and Furrow Type*: The ridge and furrow greenhouse is a community of even span greenhouses that are joined together. This design allows for more space and light. This design employs two or more A-frame greenhouses connected

Fig. 9.2 Lean to type greenhouse

Fig. 9.3 Even-span type greenhouse

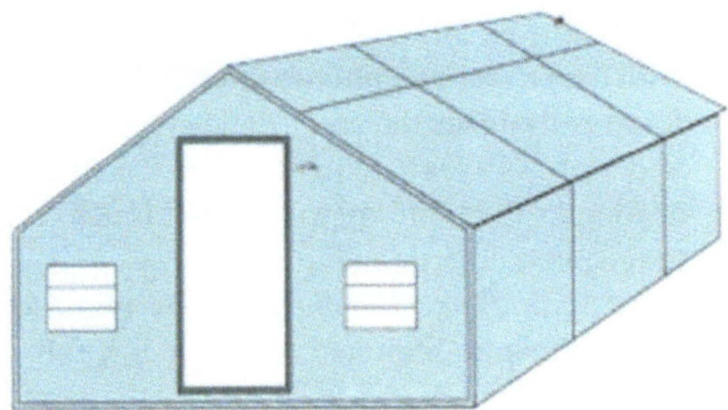

Fig. 9.4 Uneven-span type greenhouse

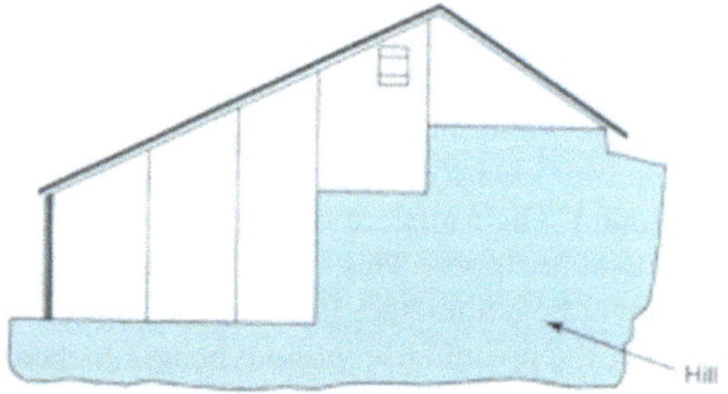

Fig. 9.5 Ridge and furrow
type greenhouse

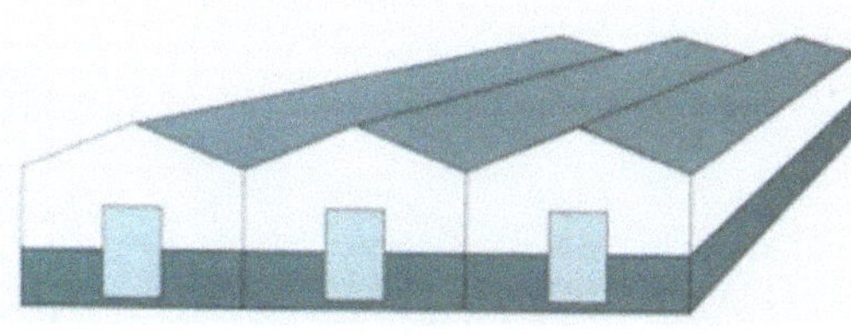

Fig. 9.6 Saw-tooth type
greenhouse

along the length of the eave. The eave acts as a furrow or gutter, carrying rain and melting snow away. The side wall between the greenhouses is removed, resulting in a structure with a single large interior. Interior space consolidation reduces labor, lowers the cost of automation, improves personal management, and reduces fuel consumption by reducing exposed wall area through which heat escapes (Fig. 9.5).

- *Saw-Tooth Type Greenhouse*: This type of greenhouses is comparable to ridge and furrow types, with the exception that this type offers natural ventilation through the greenhouse's saw-tooth shape. The saw tooth vent can be left open to improve the climate control of the growing area or closed to reduce interior temperature. Twenty-five of the covered area's overall ventilation is provided by the roof ventilation alone in addition to the side ventilation. Excellent light transmission is made possible by the arches' design (Fig. 9.6).
- *Quonset Greenhouse*: In this greenhouse, pipe purling that runs the entire length of the structure supports the pipe arches or trusses. It's shaped like a dome. Polyethylene is typically used as a covering material. In addition to being practical for a small, remote cultural area, this greenhouse is also more affordable than greenhouses that are connected to gutters. These homes are joined together either in a free-standing design or in a ridge-and-furrow pattern. The truss members of the interlocking type overlap sufficiently to permit a garden to grow in between the overlapping sections of neighboring homes (Fig. 9.7).

Moreover, greenhouses can be also classified based on working principles as active and passive greenhouse (Tiwari, 2003):

- *Active Greenhouse*: These greenhouses have access to external thermal energy, which can come from either fossil fuels or solar energy that is fed inside the greenhouse through a collector panel. To move the working fluid in the system in these greenhouses, mechanical energy is used in the form of fans and pumps.

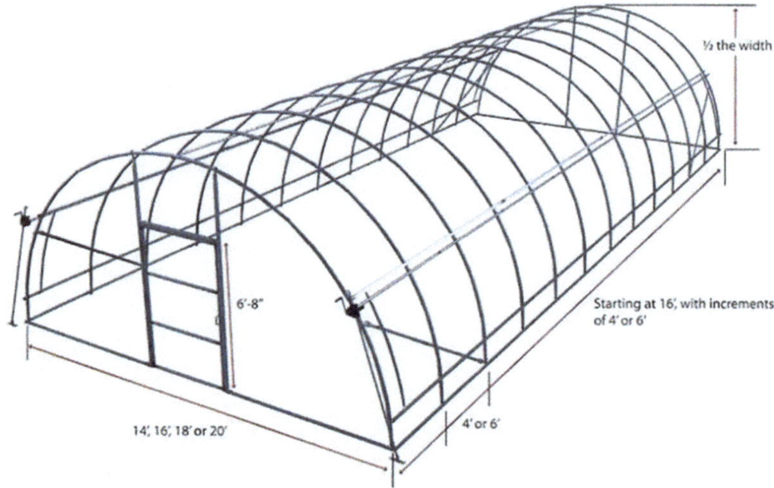

Fig. 9.7 Quonset type greenhouse

- *Passive Greenhouse*: These systems are typically those that operate without the need for mechanical energy to move fluids. The temperature gradients created by the absorption of radiation, fluids and energy move. The glazed area, walls, and roofs of the greenhouse are used to collect, store, and distribute solar energy through convection, radiation, and conduction in order to function as a collector. To reduce heating (and cooling) loads, passive systems rely solely on architectural design that can be used to maximize solar gain in the winter and minimize it in the summer.

9.4.3   Design and Construction of a Greenhouse

A greenhouse enables the maintenance of a microclimate beneficial to crops that are being considered for growth. For a particular design, its structure should be both light and strong. Greenhouses are made to last for 25 years and are regarded as semipermanent structures. A greenhouse's structure must be able to support loads like its own weight, wind, and snow, as well as transmit the most sunlight possible (Tiwari, 2003). They shouldn't be built for a life cycle of less than 10 years if the intelligent greenhouse contains high-value crops and equipment.

- *Site Selection*: In order to lower the cost of grading, the building site should be as level as possible. It should also be well-aerated and receive adequate sunlight. Because greenhouse operations require a lot of water, it is always advisable to install a drainage system. Installing tile drainage below the surface before building a greenhouse is advised in areas where drainage is a concern. A natural windbreak should be considered when choosing a site. The north and northwest sides have something, like a tree line or hill. To keep drifts away from the greenhouses in regions where snowfall is predicted, trees should be placed 30.5 m away.

- *Orientation*: According to Tiwari (2003), two general criteria for greenhouse orientation include the following: The prevailing winds should not adversely affect the facility's operation or construction, and the light level in the greenhouse should be sufficient and uniform for crop growth. In order to allow low-angle light from the winter sun to enter along a side rather than from an end where it would be blocked by the frame trusses, single greenhouses (free standing and single span) located above 40 N latitude in the northern hemisphere should be built with the ridge running East to West. Since the angle of the sun is much higher below 40°N latitude, the ridge of individual greenhouses should be oriented from north to south. At all latitudes, multi-span greenhouses or gutter-connected greenhouses (greenhouses connected to one another along their length) should be oriented north to south. In a north-south configuration, a greenhouse is spared the shadow cast by the greenhouse directly to its south in an east-west configuration.

- *Design*: The protection of the crop against harmful weather conditions (low and high temperatures, snow, hail, rain, and wind), diseases, and pests is the primary purpose of the greenhouse structure and its covering. It is crucial to design greenhouses with as much natural light as possible inside. It is important to reduce the number of structural elements that could create shadows inside the greenhouse. Therefore, the covering materials should have the most unsupported area possible and, as a result, provide the highest level of light transmission. Based on the types of frames, various greenhouse structural designs are available. The most typical shape for a greenhouse is probably one with straight side walls and an arched roof, though the gable roof is also very popular. The most common material used to build greenhouses with an arch roof and hoops is galvanized iron pipe that has been formed using a roller pipe bender.

- Low-growing crops like lettuce are suitable for hoop-style greenhouses. The best way to ensure the best operational use of the greenhouse is to use a straight side wall structure rather than a hoop-style house when growing tall growing crops or when benches are being used.

- *Selection of Equipment's for Greenhouse*: Growers in greenhouses work to maximize their output from a small, covered area. Cultivators invest in inputs to help them reach their objectives and make the most of their knowledge of how to use a variety of tools to prevent costly errors during crop production. The necessary tools/instruments for measuring various environmental parameters include thermometers, humidity meters, pH meters, and Lux meters. Commercially available portable CO_2 meters can measure the level of carbon dioxide. In the polyhouse, the level of carbon dioxide typically peaks in the morning at around 1000 ppm and then gradually decreases throughout the day. Increased levels of carbon dioxide contribute to higher net photosynthetic rates. In naturally ventilated greenhouses, opening side curtains aids in the natural replacement of reduced carbon dioxide levels.

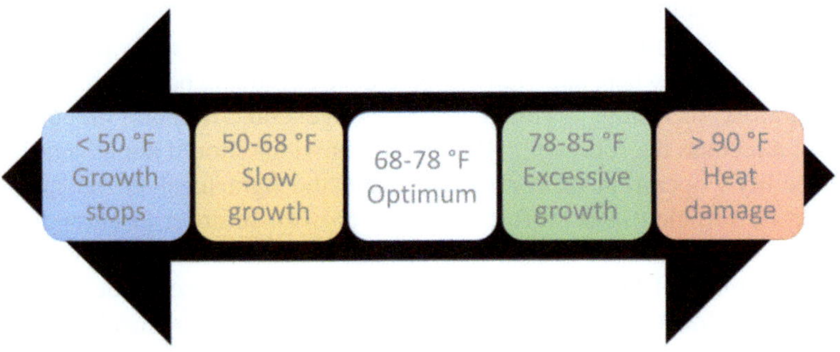

Fig. 9.8 Optimal temperature for growth

9.4.4 Greenhouse Heating and Cooling (Fig. 9.8)

9.4.4.1 Greenhouse Cooling

Cooling of the greenhouse refers to reducing the heat or hot air or protecting the crop from heat inside the structure and the quantum heat to be removed from the greenhouse is referred to the cooling load.

9.4.4.1.1 Need for Cooling System

- Every time the temperature in the greenhouse exceeds what the plants can tolerate, a cooling system is required.
- In our nation's tropical and subtropical regions, particularly the plains and coastal regions, greenhouses require cooling during the hot summer months.
- The greenhouse's temperature will always be 4–10 °C higher than the outside temperature.

9.4.4.1.2 Cooling Systems

When it comes to cooling during the autumn and spring seasons, ventilation can be used, but when it comes to cooling during the summer, other strategies must be used. Greenhouse cooling is aided by roof shading. By coating the glazing directly with opaque materials, or by covering it with aluminum or wood, the amount of solar radiant energy that enters the greenhouse can be decreased. The preferred option for this purpose is commercial shading compounds or mixtures made with paint pigments. The best materials for reflecting sunlight are white ones, which reflect 83% of it as opposed to 43% for green and 25% for blue or purple.

- *Natural Ventilation*: Some greenhouses have side and ridge vents that run the entire length of the structure and can be opened as needed to maintain the desired

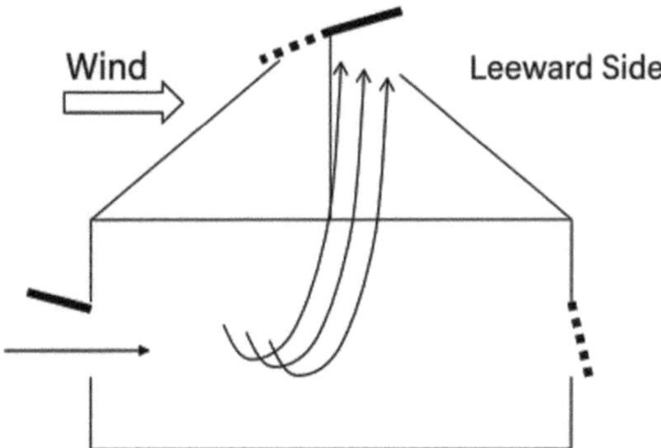

Fig. 9.9 Natural ventilation in a greenhouse

temperature. This technique makes use of thermal gradients, which cause circulation as warm air rises. Houses with only side vents are typically unsatisfactory because they rely on wind pressure for ventilation. The ridge vent must be left open so that warm air can rise through it as cooler air enters from the sides. The vent's size is crucial. The size of the side vents and ridge vents should be approximately equal to one-fourth of the floor area. To provide a roughly 60-degree angle to the roof, the roof vents should open above the horizontal position. The majority of these vents are operated by hand (Fig. 9.9).

- *Evaporative Cooling (EC)*: The amount of wet-bulb depression that takes place under a specific set of climatic conditions directly correlates to the cooling achieved by an evaporative system. Areas with consistently low relative humidity are best suited for EC systems. It can be achieved by following methods.

  1. *Fan and Pad System*: It can be used for both large and small greenhouses because it draws air through wet fibrous pads mounted on the end wall or the opposing side of the greenhouse. It uses both a vertical and a horizontal pad system. Vertical pads build up salts, which makes an opening for hot air to enter the greenhouse. The pad can be made from a variety of materials, including gravel, bark, straw, burlap (jute), aspen (a tree), wood, fiber, honeycomb, paper, etc. Volcanic rock and pumice are said to work extremely well. The majority of greenhouses need 20 m between the fan and the pad and a pad height of 1 m. 6 L of water should be available per minute for every linear meter of pad length, according to the pad design system. Place the exhaust fan farther apart than 9 m.

     Maintenance of Fan and Pad System

- (1) Because the pad cannot be frequently replaced, offer shade to block out direct sunlight.
- (2) Allow pads to dry out every day to prevent nutrient contamination, which causes holes to form when salt is deposited.
- (3) To prevent the growth of algae, clean the hollow depression with a 1% solution of sodium hypochlorite by combining the solution with water once or twice a year.

Fig. 9.10 Exhaust fan

(4) If the pads are found to be damaged and clogged with salt, they can be
    replaced.

2. *Low-Pressure Mist System*: When compared to natural ventilation, misting in
   a greenhouse with water pressure less than 7 kg/cm^2 results in air that is 5 °C
   cooler. Large and slow-evaporating water droplets are produced by low-
   pressure misting systems. This method has a significant disadvantage in that
   nutrients are leached from the soil and foliage.
3. *High-Pressure Mist System*: Low-capacity nozzles (1.8–2.8 L/h) spray water
   into the air above the plants at pressures of 35–70 kg/cm^2. Even though the
   majority of the mist evaporates before it touches the plants, some of it settles
   on the foliage and lowers leaf temperatures (Figs. 9.10 and 9.11).

9.4.4.2   Greenhouse Heating

One of the most crucial components of a successful plant production is a good heating system. Any heating system that controls temperature uniformly while not
releasing substances that are harmful to plants is acceptable. Natural gas, LP gas,
fuel oil, wood, and electricity are acceptable energy sources. The size of the greenhouse heater is determined by how much heat is lost from the building. Conduction,
convection, and radiation are the three main ways that heat is transferred out of a
greenhouse (Figs. 9.12).

9.4.4.2.1   Need for Heating

Heating is used to maintain the greenhouse at the right humidity level and at the
right temperatures for crop growth. Not just in the winter but also all year round,
heating may be necessary. In the greenhouse, heat should ideally be applied as low

Fig. 9.11 Cooling pad

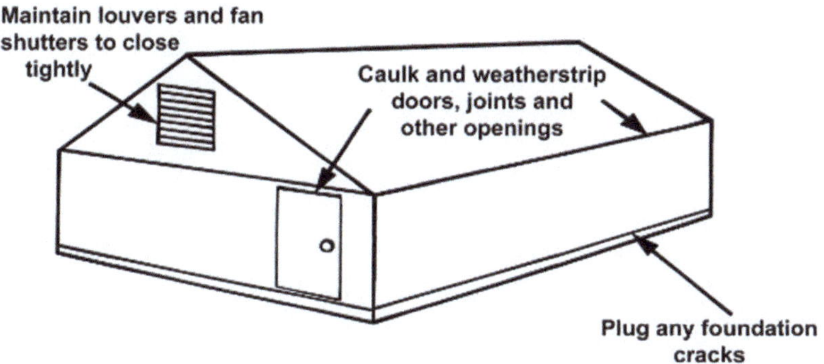

Fig. 9.12 Heat loss due to infiltration in a greenhouse. (John Worley et al., 2011)

as possible (with the exception of "grow pipes") and even heat distribution is crucial for the best crop growth.

9.4.4.2.2 Heating Systems

- *Water or Steam Heating*: In Holland, this type of heating method is very popular and is a very efficient heating system. By placing heating pipes inside the greenhouse at regular intervals, heat is directly radiated into the plants. In order to maintain a controlled, uniform temperature throughout the greenhouse, hot water is circulated through it using insulated pipes. Diesel fuel, diesel oil, gas, or electrical coils are used to power the heaters, and any of the units can be operated manually or with the aid of a control computer system.
- *Unit Heaters*: To distribute heat evenly throughout the greenhouse, a number of unit heaters at a height of 3 m must be provided as part of this localized heating

Fig. 9.13 Gravity vented unit heaters. (Scott Sanford, 2011)

system. With the aid of a fan and jet system, perforated polythene tubing running through the center of the greenhouse transfers the heat produced (Fig. 9.13).

- *Hot Air System of Heating*: As it is drawn into the cavity of the greenhouse, cold air flows over the burning coil. The heated air was then released through sleeves into the greenhouse. A centrifugal blower powered by an electrical blower completes the air-circulation process.
- *Solar Heating*: The solar heating system is the fourth potential type of system, but it is still too expensive to be a practical choice. Hobby greenhouses and small businesses both use solar heating systems. Solar energy is combined with energy storage systems made of both water and rocks. Solar heating systems are expensive, which deters the greenhouse industries from using them extensively. Water is heated during the day using flat plate solar heaters. In the insulated tanks, the hot water is kept. During the night, the hot water is circulated in the pipe that has been provided along the length of the greenhouse. On cloudy or rainy days, an additional set of emergency heating systems is available to heat the greenhouse (Fig. 9.14).
- *Radiant Heating System*: In contrast to warm air systems (like forced air unit heaters), radiant heaters direct the heat source to the floor level rather than the ceiling. Vacuum (negative) or power (positive) pressure are the two ways that hot gases are forced through the radiant tube. The reflectors positioned above the radiant tubes then direct the generated radiant energy downward. The largest mass in a building is usually the floor. As a result, the floor becomes the main heat source. A radiant heating system powered by gas imitates the sun's radiant output. The radiant tube emits radiation in all directions, just like the sun. Without a reflector, a radiant tube suffers significant convection losses. The radiant energy is directed toward the floor area by reflectors that are positioned above the radiant tube. When objects in its path absorb the radiant energy, it is converted into heat. The heat sinks in the building, such as the concrete floors, equipment, and fixtures, absorb the radiant energy. The "warmth" that is perceived in the surrounding air is re-radiated energy from this heat sink.

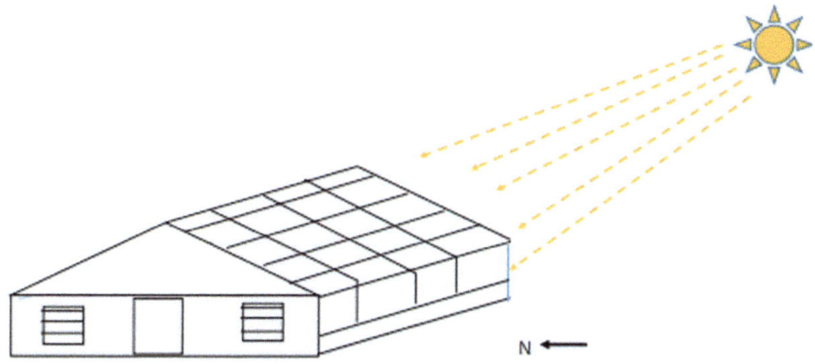

Fig. 9.14 Solar heating in a greenhouse

9.4.5   Cultivation Process in a Greenhouse

We need to take into account the following factors for successful greenhouse cultivation before beginning.

- The ideal soil pH ranges from 5.5 to 6.5, and the ideal soil EC ranges from 0.3 to 0.5 mm/cm.
- A constant supply of high-quality water is available.
- The pH and EC of the irrigation water samples should be between 5.5 and 7.0, respectively.
- The location should be free of pollution.
- Roads are necessary for market shipping and transportation of goods.
- The space ought to be adequate for the impending expansion.
- Workers should be readily available.
- There should be top-notch communication tools available.
- The soil should have excellent drainage.

9.4.5.1   Selection of Crops for a Greenhouse

Greenhouse farming requires a significant investment; thus, despite the crop's high commercial value and long-term market demand, it is grown in a greenhouse.

Floriculture crops, such as cut flowers (gerbera, carnations, orchids, and lily); vegetables such as cucumber and tomato; exotic vegetables; and fruits such as strawberry are primarily grown in greenhouses.

9.4.5.2 Cultural Practices

Different plants have different cultural and management practices. Therefore, we need to follow proper guidelines for fertilization, sterilization, irrigation, disease and pest control, etc. depending upon the crop that we are cultivating. Following are the common cultural practices that we need to follow for successful greenhouse cultivation.

9.4.5.3 Soil Sterilization

Among the nursery diseases, damping off is most prevalent and destructive disease and is caused by soil fungal pathogens like *Fusarium*, *Phytophthora,* and *Pythium.* Therefore, it is absolutely important to sterilize the soil before plantation of crops. Various methods of soil sterilization are given below:

- Steam sterilization for the conditions in India, steam sterilization, is not practical.
- Sun sterilization: Plastic sheet is placed on the ground and exposed to the sun for 6–8 weeks to sterilize it. The most fungus will be killed when the sun rises because it will heat the soil.
- Chemical sterilization: The most innovative and practical method is chemical sterilization. Silver and hydrogen peroxide (H_2O_2) are used to sterilize soil.

9.4.5.4 Media Preparation

The soil mixtures used in greenhouses to grow cut flowers and potted plants are highly modified combinations of soil and other organic and inorganic materials. Top soil is typically mixed with other ingredients to increase the potting soil's water holding capacity and aeration when it is a part of the mixture. Instead of adding topsoil to their soil mixtures, many greenhouses create their own soil mixtures out of a combination of these organic and inorganic materials. These artificial mixtures can produce crops that are just as productive as those grown in top soil when properly managed in terms of watering and fertilization practices.

The media used in greenhouses typically differ from field soils in terms of their physical and chemical characteristics.

- A desirable medium should have a good balance of physical characteristics, such as porosity and water holding capacity.
- The medium needs to be thoroughly drained.
- A medium that is too compact will have drainage and aeration issues, which will stunt root growth and possibly harbor pathogens.
- Low water and nutrient holding capacity of highly porous media will affect plant growth and development.
- For the majority of greenhouse crops, the media reaction—which has a pH range of 5.0 to 7.0 and a soluble salt (EC) level of 0.4 to 1.4 ds/m—is ideal.

- When the pH of the media is low (5.0), micronutrients like iron, zinc, manganese, and copper become toxic, and there is a deficiency in major and secondary nutrients. When the pH is high (>7.5), micronutrients like boron become deficient.

9.4.5.5 Pasteurization of Greenhouse Plant Growing Media

Greenhouse growing medium should be decontaminated by heat treatment or by treating with volatile chemicals such as methyl bromide and chloropicrin because it may contain harmful disease-causing organisms, nematodes, insects, and weed seeds.

9.4.5.6 Fertilization

One of the most important aspects of growing crops in a greenhouse is ensuring adequate nutrition. With so many variables influencing plant nutrition and an ever-growing selection of fertilizers and fertilizer systems, growers have adopted the pre- and post-plant fertilization programs listed below.

- Preplant fertilization: Preplant fertilization programs, as their name suggests, are started before crops are planted. Pre-sowing fertilization's main goal is to ensure that a sowing layer contains a balance amount of nutrients. In other words, pre-sowing fertilization aims to offer the vital nutrients needed for the early stages of crop development. Pre-sowing fertilization has a direct impact on germination and root growth in this regard.
- Post plant fertilization: After plants are transplanted, post-plant fertilization programs are started. Liquid fertilization using a water-soluble fertilizer is the most typical kind of post-plant fertilization for greenhouse crops (WSF). Because nitrogen and potassium quickly leach from the media, it is critical to maintain a steady supply. Despite being mixed in with the media, trace amounts of phosphorus and sulfur are frequently present as well. It is challenging to predict when the levels of these nutrients will be insufficient for crop needs because they may quickly leach from media
- Fertilization frequency: Additionally, fertilizer application frequency affects how quickly plants grow. Today, growers typically employ two techniques: (1) constant liquid fertilization, also known as fertigation, which calls for applying a diluted fertilizer solution at each irrigation, and (2) scheduled fertilization, which calls for applying a more concentrated fertilizer solution at regular intervals, typically once per week, and irrigating with clear water in between feedings. In some circumstances, it is crucial to provide nutrients when vegetative or reproductive growth is at its strongest.
- Fertigation system: An automatic mixing and dispensing unit with three systems pumps and a supplying device is installed in the fertigation system. The fertilizers are mixed in a specific ratio after being separately dissolved in tanks, and drippers are used to apply the mixture to the plants.

9.4.5.7   Irrigation System in a Greenhouse

A properly planned irrigation system will provide the exact amount of water required every day, all year long. The growing area, the crop, the time of year, the weather, and whether the heating or ventilation system is in use will all affect how much water is required. The type of soil or soil mixture, as well as the size and type of the container or bed, all affect how much water is required. There are many different irrigation water application systems in use today, including drip irrigation, overhead sprinklers, boom watering, perimeter watering, and overhead sprinklers.

- *Manual/Hand Watering*: It is the most popular but most expensive method of watering, and it is still used in places where labor is inexpensive, the size of the business is relatively small, and automation is not used. It takes a lot of time. However, it is still practiced in areas with dense populations of crops, such as nurseries, seed flats, or pots. Care should be taken to break the force of the water where it is being used, either by using a fine rose spray or a pipe breaker, to avoid washing the root medium out of the pots and disturbing the structure of the root medium surface.
- *Overhead Irrigation System*: The easiest and least expensive way to irrigate GH crops is from above. The pipes in this system are suspended between 60 and 180 cm above the plants. The pipes have nozzles installed in a 360° pattern. In order to prevent these nozzles from becoming clogged, care should be taken to ensure that the water is of good quality and is properly filtered (Fig. 9.15).
- *Perimeter Irrigation*: Crops grown in benches or beds can use perimeter irrigation systems. A bench's perimeter is typically encircled by a plastic pipe with water-spraying nozzles that spray water over the substrate surface beneath the foliage. You can use PVC pipe or polythene. While polythene pipe tends to roll if it is not firmly fastened to the side of the bench, PVC pipe has the benefit of being very stationary. As a result, nozzles move out of alignment with the surface

Fig. 9.15 Overhead irrigation in a greenhouse (grow span)

of the substrate. Nozzles are available to put out a spray at an angle of 180°, 90°, or 45° and are made of nylon or a hard plastic. No matter what kind of nozzles are employed, they are spaced evenly across the benches so that each one extends between two others on the opposite side.

- *Drip Irrigation*: This irrigation system provides the plant with water in proportion to its consumptive needs. This irrigation system uses very little water and has a very low water requirement. The system automatically handles the limited water supply as the water is applied through drippers. The four fundamental portions of a drip system are suction, regulation, control, and discharge, which are carried out by a water lifting pump, hydro cyclone filter, sand filter, fertilizer mixing tank, screen filter, pressure regulator, water meter, main line, laterals, and drippers. In order to supply water with pressure, a water lifting pump is necessary. The water is lifted, then filtered by a hydrocyclone, a sand filter, a tank for mixing fertilizer, a screen filter, and finally a dripper. Relatively coarser particles are filtered through hydro cyclone filters, finer particles are filtered through sand filters, and very fine particles are filtered through screen filters. These filters are necessary for the water to flow through drippers and laterals without clogging; otherwise, this could happen (Fig. 9.16).

Drip Systems Come in Two Different Types
- The high-pressure drip system operates at a pressure of at least 30 psi.
- Low-pressure drip system: This system functions at a pressure of less than 30 psi.

Fig. 9.16 Drip irrigation in a greenhouse (Idromeccanica Lucchini)

MCQs

1. Which of the following cannot be categorized as a method of protected farming?

   - Mulching
   - Greenhouse cultivation
   - Shade net houses
   - None of the above

2. Agriculture production within protected structures began in the nineteenth century

   - France and Netherlands
   - Israel
   - Germany
   - Japan

3. Which of the following greenhouse is suitable for hillside slopes?

   - Even span green house
   - Lean to type greenhouse
   - Uneven span green house
   - Saw tooth type greenhouse

4. What is the ideal range of soil pH for greenhouse cultivation?

   - 4.5–5
   - 5.5–6.5
   - 5–7
   - 6.5–8

5. Evaporative cooling can be achieved by

   - High-pressure mist system
   - Radiant heating system
   - Fan and pad system
   - Both 1 and 3

6. The high-pressure drip system operates at a pressure of at least

   - 30 psi
   - 45 psi
   - 60 psi
   - 25 psi

7. In a drip irrigation system, relatively coarser particles are filtered through

   - Sand filters
   - Screen filters
   - Hydrocyclone filters
   - None of the above

8. Which of the following is suitable for growing in a greenhouse?

- Carnation
- Orchids
- Tomato
- All of the above

9. ______ Have walls that are at least 4 m high and have roof peaks that are up to 8 m high.

- Low-technology greenhouse
- Medium-technology greenhouse
- High-Technology greenhouse
- None of the above

10. At all latitudes, multi-span greenhouses or gutter-connected greenhouses should be oriented

- East West
- North South
- South West
- East North

References

Cedillo, E., & Calzada, M. (2012). *La Horticultura Protegida en Mexico Situacion Actual Perspectivas* (pp. 1–10). Encuentros UNAM.

Food and Agriculture Organization of the United Nations. (2020). *High level expert forum – How to feed the world in 2050?* Office of the Director, Agricultural Development Economics Division.

Keshwania, N., Vishal, & Nanglia, S. (2021). *Protected cultivation introduction, 3, 4.* Just Agriculture Pvt Ltd

Pini, M., Marucco, G., Falco, G., Nicola, M., & Wilde, W. (2020). *Experimental testbed and methodology for the assessment of RTK GNSS receivers used in precision agriculture.*

Protected cultivation & post harvest technology, Course developed by Tamil Nadu Agricultural University, 41.

Sanford, S. (2011). *Greenhouse unit heaters, types, placements & efficiency*

Tiwari, G. (2003). *Greenhouse technology for controlled environment.* Alpha Science International Ltd..

Worley, J., et al. (2011). Greenhouse *heating, cooling and ventilation.* Bulletin 796. The University of Georgia and Ft. Valley State University, the U.S. Department of Agriculture and counties of the state cooperating.

Chapter 10
Organic Farming and Sustainable Agriculture

Abstract In recent years, organic farming is gaining increasing popularity due to growing health and environmental awareness among people. Indirectly, it has proved to be a boon for Indian agricultural scenario; therefore, due attention has been given to it in the introductory sections of this chapter. Organic foods are preferred by people because they are grown without any use of synthetic inputs and hence pose no risk to one's health, whatsoever. This chapter throws light on the scope and objectives of organic farming, farming systems, and cropping patterns. A section about sustainable agriculture has also been given toward the very end of chapter with topics of higher gravity like HEIA, LEIA, LIESA and IFS, discussed in detail.

Keywords Organic · Pesticide residues · Sustainability · Cropping Pattern · Cropping Scheme · Cropping System · HEIA · LEIA · LIESA · IFS

10.1 Introduction and Definition of Organic Farming

If looked upon fundamentally, organic farming is an agricultural practice in which fertilizers and pest control tools are made from or obtained from organic elements. Such a mechanism prevents our ecosystem from being harmed and degraded.

There are various technical definitions of organic farming put forward by various agricultural organizations like FAO and IFOAM. They are mentioned below:

According to IFOAM, "Organic Agriculture is a production system that sustains the health of soils, ecosystems, and people. It relies on ecological processes, biodiversity and cycles adapted to local conditions, rather than the use of inputs with adverse effects. Organic Agriculture combines tradition, innovation, and science to benefit the shared environment and promote fair relationships and good quality of life for all involved" (IFOAM, 2008).

According to FAO, "Organic agriculture is a holistic production management system which promotes and enhances agro-ecosystem health, including biodiversity, biological cycles, and soil biological activity. It emphasizes the use of management practices in preference to the use of off-farm inputs, taking into account that regional conditions require locally adapted systems. This is accomplished by using, where possible, agronomic, biological, and mechanical methods, as opposed to using synthetic materials, to fulfil any specific function within the system" (FAO/WHO Codex Alimentarius Commission, 1999).

10.2   Scope of Organic Farming in India

Scope of organic farming in a country like India is immense. All kinds of crops thrive in India due to the diverse agroclimatic zones and cropping situations. Because of India's massive agriculture economy, the potential for creating, storing, and utilizing nutrient packed on-farm inputs, such as crop residues, organic manures, green manures, and biopesticides, is enormous. When compared to developed countries, India's 70% rural population is impregnated with indigenous technical knowledge (ITK) and skills, as well as the availability of skilled and trained manpower at significantly lower cost.

Organic food production costs more in developed countries because organic farming is labor-intensive, and labor is expensive in these countries. However, in a country like India, labor is plentiful and cheap. OF is viewed as a good cost-effective solution to the rising costs of chemical farming.

The strategic placement of India on world map also plays an important role in scope and future growth prospects of organic farming. India lies in between the western developed countries like the United Kingdom and the United States and countries like Singapore on the East. This strategic placement of India on the world map makes it easier to export products of organic farming to these countries. Along with this, there has been a significant rise in demand if organic products like vegetables and dairy products. This has been seen as a direct correlation with the rising health consciousness in the west. This demand is also seeing an exponential growth in the domestic markets. Also, the Government of India has always been in constant support and an advocate for organic farming with introduction of government schemes in favor for the same.

10.3   Objectives of Organic Farming

- To sustain and increase the soil fertility
- To cooperate with natural systems as opposed to attempting to deteriorate them
- To reduce the pollution caused by agriculture and allied activities especially the rise of plasticulture, excessive use of chemicals, etc.
- To give farmers a sufficient return on their investment and a sense of fulfilment from their labor, alongside equal opportunities at life and improved food and nutritional security
- To produce adequate quantities of nutritionally rich food

10.4 Choice of Crops for Organic Farming

- Before procuring seeds, one must make sure that they are certified "organic."
- It is prohibited to use genetically modified seeds, transgenic plants, or plant material when it comes to organic farming (*APEDA*).
- Chemically untreated conventional seed and plant material must be used if certified organic seed and plant materials are not readily accessible.

10.5 Introduction to Farming Systems

The farming system is a decision-making tool and land-use unit that includes the farm household, cropping, and livestock systems, which convert land, capital, and labor into marketable products that can be sold.

Farming system is a suitable combination of farm businesses such as cropping and livestock system, horticulture, fishery, poultry, forestry, and the means available to farmers to increase their profitability.

It interacts with the environment without disrupting the ecological and socioeconomic balance on the one hand while attempting to achieve the national objective of food, fiber, fodder, and fuel on the other.

10.6 Concepts in a Farming System

- In a farming system, a farm is referred to as a whole. It includes agriculture, horticulture, sericulture, beekeeping, livestock rearing, poultry, forestry, etc. The approach we take in this case is always holistic.
- The farm should be viewed as a whole and planned for the efficient integration of the businesses that will be combined with crop production, ensuring that the waste products and by-products of one business are successfully used as inputs in another.
- When carefully selected, planned, and carried out, a combination of one or more crop-related businesses can yield greater dividends than a single business, particularly for small and marginal farmers (Fig. 10.1).

10.7 Components of a Farming System

Majorly, a farming system is consisting of following components:

1. Cropping
2. Dairy farming
3. Goat farming

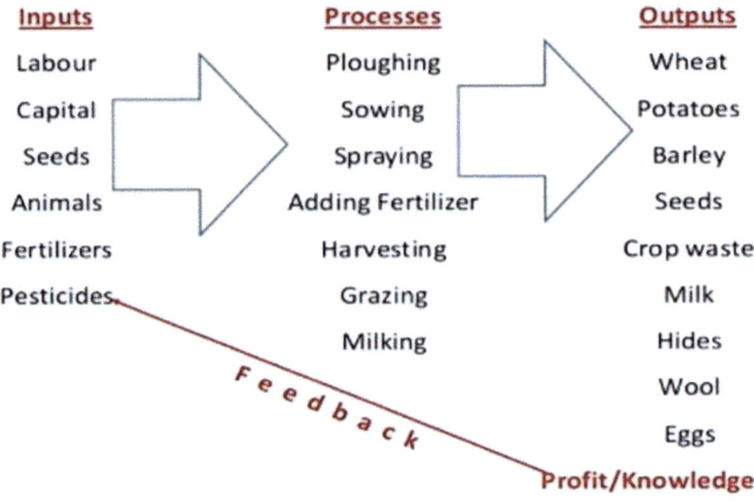

Fig. 10.1 Flow chart of activities in a farming system

  4. Sheep farming
  5. Pig farming/piggery
  6. Rabbit farming
  7. Camel farming
  8. Poultry farming
  9. Duck farming
 10. Turkey farming
 11. Fish farming/fishery
 12. Prawn farming
 13. Sericulture
 14. Lac farming
 15. Apiculture
 16. Mushroom cultivation
 17. Horticulture
 18. Agroforestry
 19. Vermicomposting
 20. Biogasplant

10.8   Cropping Pattern

The proportion of farmland planted with various crops at various times is referred to as the cropping pattern. It includes the seasonal order and geographic arrangement of crops and fallow in a particular region. Cropping patterns change over time and place as a result of rainfall, climate, temperature, soil type, and automation.

10.8.1 Factors Affecting Cropping Pattern

- Availability of inputs
- Change or modification in technology
- Fluctuations in any major agriculture policy of the government

10.9 Cropping System

The term "cropping system" applies to both the cropping pattern and how it interacts with the environment and other resources.

Consequently, a cropping system entails a cropping plan in addition to all the elements needed to produce a specific crop, as well as the interactions between those elements and the environment. The management and organization of crops is a key component of creating an efficient ecological agricultural system that makes the best use of the resources at hand (soil, air, light from the sun, water, labor, and tools). It depicts the types of crops grown on a farm and how they interact with other farm supplies, farm businesses, and technological advancements to form their structure. It is put into action at the ground level.

Since it is location-specific, it alters as place and surroundings do.

10.9.1 Types of Cropping Systems

10.9.1.1 Mono-cropping Systems

It is a method of growing the same crop or just one crop every year on the same piece of land. It is also known as single cropping or monoculture. Therefore, cropping intensity is always 100%.

10.9.1.2 Multiple Cropping Systems

It is described as growing two or more crops in a row on the same area without lowering soil fertility. It is an uptick in cropping in both temporal and space dimensions. Thus, there are more crops produced within a year on a given area of land during a specific growing season.

It is mainly of four types:

1. *Intercropping*: Growing two or more crops on a single piece of land in a scientifically sound manner, in definite rows with proper arrangement. The land use efficiency increases by practicing this. It is further of many different types, alley, strip, contour cropping, etc.

2. *Mixed Cropping*: In this type, we witness the cultivation of two or more crops on a same acreage of land, but this method is unscientific. There is no fixed arrangement, and the broadcasting method is taken use of. The main reason of performing this is to lessen the risk of crop failure. Even if one fails, other one will survive.
3. *Sequential Cropping*: It can be summed up as growing two or more crops quickly after one another on the same plot of ground during a crop season. It indicates that the following crop will be sown soon after the previous crop is harvested, or even concurrently. Nonoverlapping cropping is another name for this.
4. *Relay Cropping*: Growing two or more crops at once throughout each crop's life span. Before the previous crop is harvested, the succeeding crop is sown. It indicates that the second crop is sown after the first has reached the reproductive stage of development but before it is prepared for harvest.

10.10   Cropping Scheme

Cropping scheme refers to the method by which crops are grown on separate field areas in order to maximize the return from cash crops while maintaining the fertility of the land.

10.10.1   Importance of Cropping Scheme

- It supports effective practical planning.
- It aids in figuring out real net profit.
- It depicts the farm's typical and abnormal circumstances.
- It enables the estimation of potential crop profit or loss.
- It displays the price of periodic and varietal farming.
- For effective labor, capital, and land administration, it is necessary.

10.11   Sustainable Agriculture

The ability of a farm to produce food without seriously endangering the well-being of the ecosystem is referred to as sustainable agriculture.

Agriculture creates methods for cultivating crops and raising livestock that, like nature, are self-sustaining by adhering to the laws of nature.

Figure 10.2 illustrates that sustainable agriculture can only be attained when the three main objectives will be integrated into the work, that is, the environmental health, social and economic equity in the farming community, and reasonable monetary profitability.

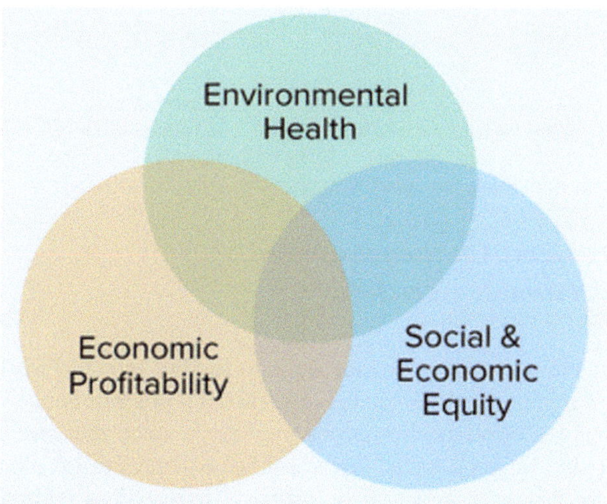

Fig. 10.2 Sustainable agriculture

A sustainable farming system can be ensured by all participants in the food system, including farmers, food manufacturers, distributors, merchants, customers, and waste administrators.

The majority of our food has been created through industrial agriculture for many years especially post 1960s; this system is characterized by large fields that cultivate the same products year after year while utilizing massive quantities of chemical pesticides and fertilizers that harm our land, water, air, and climate. Because it wastes and degrades the resources it relies on, this structure is not designed to last.

This is where the sustainable agriculture comes into the picture (Fig. 10.3).

The abovementioned figure states reasons why we need sustainability in agriculture.

10.11.1   *Practices of Sustainable Agriculture*

- Crop rotation
- Practicing no tillage
- Extensive use of integrated farming system (IFS)
- Inculcating agroforestry
- Taking full use of integrated pest management (IPM)

Fig. 10.3 Problems in modern day agriculture practices

10.11.2   Problems/Challenges Faced by Sustainable Agriculture

- Not growing enough food: Growing enough food to feed the world's expanding population is a major task for sustainable farming practitioners.
- Climate Change: Weather aberrations may see a steep increase due to climate change globally. In some areas, lower rainfall will result in prolonged droughts, which means that less land will be properly irrigated as compared to before. Leading to lower productivity.
- Shrinking of arable land: Deforestation and overfishing damage the ecosystems and wildlife, and our water supplies stand overused as of 2023. Thirty-three of the planet's territory has already suffered mild to severe degradation (FAO; 2017). The leftover land must therefore be used more effectively.
- Resources: Sustainable farming practices include safeguards for the environment, the land, and water. Therefore, it is important to lessen CO_2, methane, and nitrous oxide emissions; safeguard surface and subterranean water; and prevent erosion and detrimental compaction.

10.11.3   Indicators of Sustainability

We can recognize, measure, and assess the global impacts of sustainable agriculture with the aid of indicators. They give us a way to gauge how sustainable agriculture affects the economy, society, and ecology as well as evaluate the success of policies and the future viability of the industry.

The main indicators can be summed up as mentioned below:

- Profitability
- Land and water quality to sustain production
- Managerial skills
- Off-site environmental impacts

These indicators can highlight strengths and weaknesses and show trends in the sustainable agriculture, which can help us decide the direction to which it is headed and the appropriate measures that we need to take to set it up on the right path of sustainability with development.

10.11.4 HEIA, LEIA, LEISA, and IFS for Agricultural Sustainability

10.11.4.1 High External Input Agriculture (HEIA)

We find the roots of HEIA in the Green Revolution. It aimed at more harvest per unit area and intensification of number of cultivation seasons. It took use of high-yielding varieties, chemical fertilizers, herbicides, and pesticides, along with improved mechanization and irrigation techniques.

Unfortunately, HEIA didn't go hand in hand with sustainability. Planting a few cash crops caused the environmental equilibrium to collapse because of lesser biodiversity. Increased soil runoff as a result of repeated furrowing by machinery was one major problem. Along with this, there was huge dependence on foreign inputs such as hybrid seeds, chemical fertilizer, pesticides, and heavy machinery.

Commercial agricultural pesticides are used, which have negative effects on soil pH, soil structure, soil texture, cation exchange capacity, and soil biome in general.

The yield per unit is currently declining, and many benefits of HEIA are gradually fading away. Therefore, a sustainable agriculture system devoid of significant external inputs must be pursued by farmers, academicians, and extensionists.

10.11.4.2 Low External Input Agriculture (LEIA)

It is described as a production process that employs artificial fertilizers or other agrochemicals at levels below those typically advised. It includes maximizing the use of local resources while reducing reliance on outside inputs for agricultural production. It does not mean eliminating these supplies, but yields are maintained by a greater focus on cultural practices, IPM, INM, and on-farm resource utilization optimally.

As mentioned above, it is built on reducing synthetic inputs—not necessarily eliminating them. LEIA emphasizes the use of methods that incorporate natural processes into agricultural processes, including nutrient cycling, biological nitrogen fixation (BNF), soil regeneration, and natural enemies of pests (Pieri, 1995; Snapp et al., 1998; Kamalpreet Kaur et al., 2022).

10.11.4.3 Low External Input Sustainable Agriculture (LEISA)

In today's world, due to unavailability of labor and capital alongside lower productivity due to overexploitation of soil coupled with environment saturation of synthetic fertilizers, there is a dire need to practice Low External Input Sustainable Agriculture (LEISA).

Economically viable, ecologically sound, culturally appropriate, and socially just, LEISA makes the best use of locally available natural and human resources (such as soil, water, vegetation, local plants and animals, and human labor, knowledge, and skill) alongside recycling of plant nutrients, including nitrogen, and reducing agricultural losses brought on by insects and pests. LEISA upholds ecological standards that are compatible with indigenous ecosystems.

LEISA operates under the tenet that a favorable environment for plant development and sustenance can be created by encouraging soil microorganisms as much as possible and adding enough organic matter, encouraging biodiversity, and reducing losses from sun radiation, air, and water flows by managing the microclimate, managing the water, and controlling erosion using various biological and mechanical techniques.

Under LEISA, soil and water can be conserved with the help of procedures like mulching and no tillage cultivation. Control of weeds can be done by sowing intercrops, which doesn't give space for weed growth but also increases the land use efficiency. Enhancement of soil fertility can be done by adding manures and compost to the soil. Green manuring is also one such method used to improve the soil quality.

10.11.4.4 Integrated Farming Systems (IFS)

Integrated farming system inculcates agriculture with livestock, poultry, and fish keeping it a tightly knit system to produce year-round work and extra income. Field crops are produced for food supply in an IFS model. Horticultural and vegetable crops can also produce two to three times as much energy as cereal crops, ensuring nutritional security and revenue sustainability on the same plot of land. Crop residues can be used as livestock fodder for dairy and goat production after harvesting. Animal excreta can also be used as organic fertilizer or vermicomposting, which enhances soil fertility and thus reduces the need for chemical fertilizers. Again, animal excreta can be desiccated, composted, or liquid composted to generate biogas and energy for domestic use.

It is founded on the ideas that "there is no waste" and "waste is only a misplaced resource," which means that waste from one component becomes an input for another element of the system.

The IFS approach is regarded as the most powerful instrument for increasing the profitability of agricultural systems, particularly for small and marginal farmers, in order to make them bountiful.

10.11.4.4.1 Goals of IFS

- Enhancing productivity per unit area
- Proper waste management
- Year-round income generation
- Chemical input reduction
- Yield maximization of all component businesses
- Soil health management

10.11.4.4.2 Components of an IFS

General components of an IFS system include the following:

- Fruit cultivation
- Poultry
- Duckery
- Apiary
- Fishery
- Piggery
- Plantation crops
- Vermicomposting
- Mushroom
- Cultivation

One should keep in mind that these components can change and mix match according to the feasibility and availability.

MCQs

1. Which of the following is prohibited to use in organic farming?

   - Genetically modified seeds
   - Transgenic plants
   - Chemical fertilizers and pesticides
   - All of the above

2. What is the cropping intensity in case of mono-cropping system?

   - 100%
   - 78%
   - 10%
   - 50%

3. Which is not an objective of sustainable agriculture?

   - Maintaining a healthy environment
   - Economic profitability
   - Higher crop production through industrial agriculture
   - Social and economic equity

4. High External Input Agriculture (HEIA) does not rely on which of the following:

   - Chemical fertilizers and pesticides
   - Maximizing the use of local resources
   - Improved mechanization and irrigation techniques
   - High-yielding varieties

5. ________emphasizes the use of methods that incorporate natural processes into agricultural processes, including nutrient cycling, biological nitrogen fixation, and natural enemies of pests.

- HEIA
- LEIA
- Both 1 and 2
- None of the above

6. IFM stands for

- Integrated Farming Management
- Indian Farming Management
- International Farming Management
- All of the above

7. LEISA is related to

- Natural farming
- Organic farming
- Inorganic farming
- All of these

8. Mixed cropping is growing of

- Two or more crops simultaneously intermingled without any row pattern
- A crop intermingled with sericulture
- A crop along with apiculture
- Growing two or more crops simultaneously in definite rows

9. ______ finds its roots in the green revolution

- IFS
- HEIA
- LEISA
- LEIA

10. Which of the following is a component of IFS?

- Vermicomposting
- Mushroom cultivation
- Apiculture
- All of the above

References

FAO/WHO Codex Alimentarius Commission. (1999). *Codex alimentarius: General requirements (Food Hygiene)*. Food and Agriculture Organization of the United Nations and World. Health Organization

https://apeda.gov.in/apedawebsite/organic/organic_contents/Appendix_1_Crop%20 Production.pdf

https://www.fao.org/3/i6583e/i6583e.pdf

https://www.fao.org/organicag/oa-faq/oa-faq1/en/

https://www.ifoam.bio/why-organic/organic-landmarks/definition-organic

International Federation of Organic Agriculture Movements (IFOAM). (2008). *The IFOAM Norms for Organic Production and Processing*. IFOAM.

Kamalpreet Kaur, et al. (2022). Sustainable agriculture: Impact of LEISA and HEIA. *International Journal of Advances in Agricultural Science & Technology*, 1–8.

Pieri, C. (1995). Long-term soil management experiments in semi-arid francophone Africa. In R, Lal & B. A. Stewart (Eds.), *Soil management: Experimental basis for sustainability and environmental quality* (pp. 225–266). CRC Press.

Snapp, S.S., Mafongoya, P.L., & Waddington, S. (1998). Organic matter technologies for integrated nutrient management in smallholder cropping systems of southern Africa. *Agriculture, Ecosystems & Environment, 71*(1-3), 185–200.

Chapter 11
Agrometeorology

Abstract Agrometeorology studies the relationship between atmospheric conditions and agricultural production. It studies the weather variables that have relevance to agriculture and their effect on crop production. This part of the book starts with the importance and scope of AgroMet. Later half of it revolves around the different facets of AgroMet including clouds, condensation, weather extremes, Indian Monsoon, and weather forecasting.

Keywords Agrometeorology · Precipitation · Humidity · Dew Point · Sleet · Drizzle · Snow · Hail · Monsoon · Weather Forecast

11.1 Introduction to Agrometeorology

Agricultural meteorology is a specialized branch of applied meteorology that looks at the environmental factors affecting growing plants and animals. It is an applied science that examines the connection between weather/climatic factors and agricultural output. Agrometeorology, abbreviated from agricultural meteorology, is a science that applies meteorology to the evaluation of the physical environment in agricultural systems (Chandrasekaran et al., 2018).

According to Molga (1962), agrometeorology is the study of the meteorological, climatological, and hydrological conditions that are important to agriculture because of their interactions with the objects and processes used in agricultural production (Mavi & Tupper, 2004).

11.1.1 Scope of Agrometeorology

The key determinants of whether agriculture succeeds or fails are weather and climate. Weather has an impact on agricultural activities from crop planting to harvest, and rainfed agriculture in particular is at the mercy of the weather. Of all aspects of

L. Ahmad et al., *Fundamentals and Applications of Crop and Climate Science*,
https://doi.org/10.1007/978-3-031-61459-0_11

human life, agriculture is the most weather-dependent. Development and advancements in agricultural meteorology have been made necessary by rising climatic variability, coupled with climate change and a rise in the frequency of extreme occurrences. The study of agrometeorology is required for the following reasons:

- *Management of crops*: Numerous agricultural tasks, such as fertilizer application and planting, are involved in crop management. On the basis of carefully designed weather support, tasks like plant protection, irrigation scheduling, harvesting, etc. For this, operational projections from agro-met advisories are used. For instance, when there hasn't been any rain, the soil is 90% moist, and the wind speed is under 25 km/h, and spraying or dusting is done.
- *Crop planning for production stability*: Appropriate crops, cropping patterns, and contingent cropping planning can be chosen by taking into account the crop's water needs, rainfall, and available soil moisture. This will reduce the risk of crop failure due to climatic factors and result in yields that are stable even in the face of adversity.
- *Agricultural climate categorization*: Air temperature, rainfall, wind, relative humidity, and other climatic elements that influence crop growth, development, and yield are significant climatic factors to consider when defining crop growing season. Agro-meteorology takes into account and evaluates the suitability of these parameters in a given region for optimal crop production and economic benefits.
- *Climate extremes*: Extreme weather events such as frost floods, droughts, hail storms, and high winds can be predicted and crops can be safeguarded.
- *A tool for analyzing soil moisture stress*: The climatic water balance method can precisely measure soil moisture. This approach is used to diagnose soil moisture stress and drought, and it allows for the implementation of appropriate protective measures like irrigation, mulching, anti-transpirant application, defoliation, and thinning, among others.
- *Study crop climate relationship*: Agrometeorology investigates microclimatic aspects of crop canopy in order to modify them for increased crop growth.
- *Formation of soil*: Climate is a significant element in the formation and development of soil since it affects factors including temperature, precipitation, humidity, and wind (L. Ahmad et al., 2017).

11.2 Composition and Structure of Atmosphere

The physical gaseous mixture that envelopes earth on all sides and is tasteless, colorless, and odorless is known as the atmosphere. It is expandable, compressible, and movable. These are certain uses of atmosphere:

- Offers oxygen that is beneficial for crop respiration
- Provides CO_2 for photosynthesis, which creates biomass
- Offers N, a nutrient necessary for plant growth

- Serves as a vehicle for the movement of pollen
- Guards against damaging UV radiation for both people and crops
- Provides rain to crops grown in fields

11.2.1 Composition of Atmosphere

1. *Gases*: Nitrogen and oxygen together account for nearly 99% of the dry atmosphere's volume. The remaining 1% is made up of a variety of gases, the most vital for supporting life on Earth being ozone (0.00006%), carbon dioxide (0.03%), nitrous oxide, and methane. Ozone is concentrated in a layer that rises between 15 and 55 km above the surface of the Earth. Because it filters off the sun's harmful UV radiation, ozone is essential to life. Areas of substantial weakening have been found in the ozone layer recently, mostly at the South Pole. Researchers have discovered that the atmospheric release of the synthetic chemical chlorofluorocarbon is what is responsible for this thinning (Fig. 11.1).
2. *Water vapor*: Although it only exists in small volumes, water vapor is one of the most changeable gases in the atmosphere and is therefore crucial. The volume percentage of water vapor in air can range from 0.02% in a cold, dry region to about 4% in the wet tropics. Water vapor acts as an insulator in the atmosphere, similar to CO_2.
3. *Particulate matter*: The solid particles that make up the atmosphere include smoke, pollen, salt, dust, and volcanic ash, among others. The production of clouds and fogs is heavily influenced by dust particles. It is responsible for giving the sky its fiery red and orange hue at sunrise and sunset (Fig. 11.2).

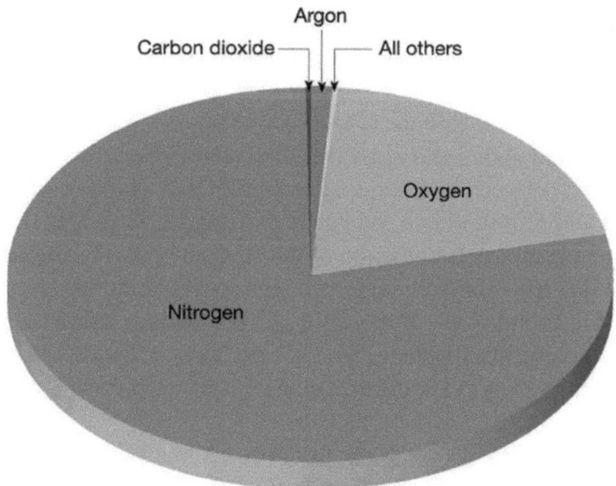

Fig. 11.1 Modern composition of atmosphere. (Fullarton, 2016)

Constituents	% by volume
Nitrogen (N_2)	78.08
Oxygen (O_2)	20.94
* Argon (Ar)	0.93
Carbon dioxide (CO_2)	0.03
* Neon (Ne)	0.0018
* Helium (He)	0.0005
Ozone (O_3)	0.00006
Hydrogen (H_2)	0.00005
* Krypton (Kr)	Trace
* Xenon (Xe)	Trace
Methane (Me)	Trace

*Inert chemically never found in any chemical compounds.

Fig. 11.2 Composition of atmosphere. (Chandrasekaran et al., 2018)

11.2.2 Structure of Atmosphere

The earth's atmosphere is divided into zones or layers that are grouped in the shape of spherical shells dependent on altitude above the earth's surface. The atmosphere is divided into the following major realms.

1. *Troposphere*: About 75% of the atmosphere's total gaseous mass is found in the troposphere. Its root is the Greek word "tropos," which means "mixing" or "turbulence." This atmosphere's lowest layer typically rises to a height of 14 km above sea level. At the poles, the troposphere is typically around 8 km high, whereas at the equator, it is roughly 16 km high. Every type of meteorological phenomenon happens in the troposphere. The temperature drops as we ascend higher into the troposphere. As we ascend, temperature drops at a rate of 6.5 °C/1000 m, which is known as the normal lapse rate. The tropopause, which separates the troposphere and stratosphere, is located at the top of the troposphere.

2. *Stratosphere*: The stratosphere is the second-lowest part of the atmosphere on Earth. It is located above and apart from the troposphere by the tropopause. This layer is located between the stratopause and the top of the troposphere, which is between 50 and 55 km above the surface of the Earth. About 1/1000 of the air pressure at sea level exists at the top of the stratosphere. The region of the Earth's atmosphere that has relatively large concentrations of that gas is called the ozone layer, and it is found there. Increasing altitude causes temperatures to rise in a layer known as the stratosphere. The ozone layer is located in the layer of the stratosphere in the Earth's atmosphere. Sunlight's UV rays are absorbed by the ozone layer.

3. *Mesosphere*: The mesosphere is the third lowest layer of the Earth's atmosphere, and it is here that meteorites burn up before they reach the Earth. The temperature drops as we ascend in this stratum. The mesosphere sits on top of the

stratosphere. The mesosphere is the third highest layer of the Earth's atmosphere, located above the stratosphere and under the thermosphere. It stretches from the stratopause at around 50 km to the mesopause at 80–85 km above sea level. Temperatures fall with increasing altitude to the mesopause, which marks the top of the atmosphere's middle layer. It is the coldest spot on Earth, with an average temperature of roughly −85 °C.

4. *Thermosphere*: One of the atmosphere's uppermost layers is the thermosphere. The ionosphere and exosphere are also included. It is above the ionosphere and below it is the exosphere. When charged particles release photons of light into the atmosphere, the phenomenon known as the Aurora Borealis appears. It stretches from the mesopause, which separates it from the mesosphere, at an altitude of about 80 km, all the way up to the thermopause, which is between 500 and 1000 km away. The thermopause is also known as the exobase due to its location at the exosphere's lower boundary. The ionosphere is located between 80 and 550 km above Earth's surface in the thermosphere's lower part. The temperature of the thermosphere continuously increments with level and can reach 1500 °C.

5. *Exosphere*: The topmost part of Earth's atmosphere is called the exosphere. From the exobase, which lies at the top of the thermosphere at a height of around 700 km above sea level, it spreads to a distance of about 10,000 km, where it combines with the solar wind. This layer is mostly made up of heavier molecules including nitrogen, oxygen, and carbon dioxide as well as hydrogen and helium, both of which have extremely low densities. The distance between the atoms and molecules allows them to move hundreds of km apart without running into one another. As a result, the exosphere ceases to function as a gas and particles continue to continually escape into space (Fig. 11.3).

11.3 Weather and Climate

11.3.1 Weather

Weather describes atmospheric conditions that occur locally over short time intervals ranging from minutes to hours or days. In other terms, weather is the physical condition of the atmosphere at a certain time. Rain, snow, clouds, winds, floods, and thunderstorms are all common examples. Weather is defined as the current state of the atmosphere at any location. It fluctuates from day to day and location to location. The following variables, known as weather elements, influence the state of the atmosphere. They are as follows: (1) air temperature, (2) pressure, (3) humidity, (4) rainfall, and (5) wind. These weather variables are inextricably linked. Most daily activities, such as farming, forms of transportation, and clothes, are influenced by the weather.

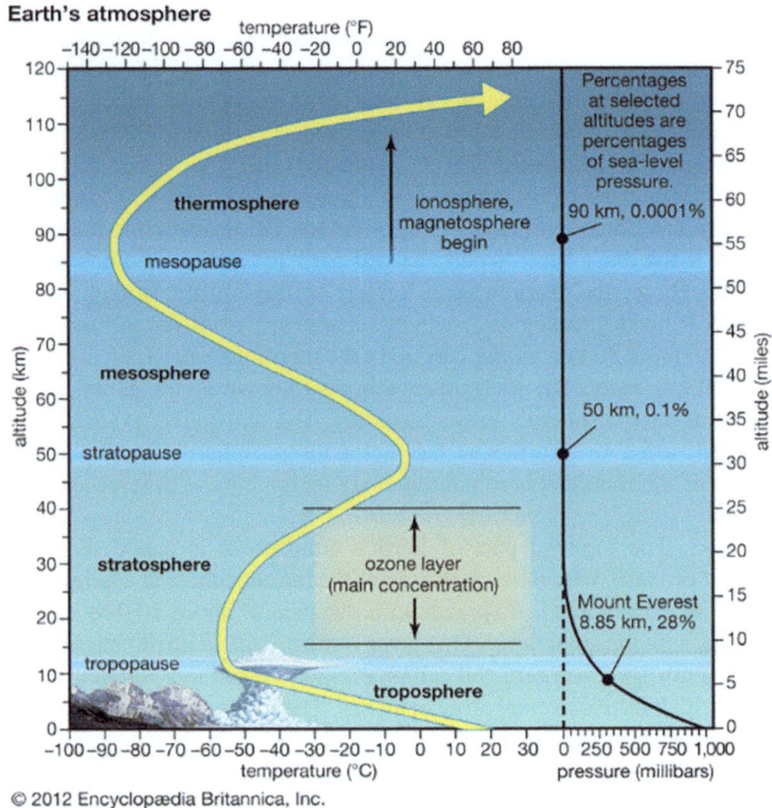

Fig. 11.3 Structure of atmosphere. (Boyer et al., 2019)

11.3.2 Climate

The long-term regional or even worldwide average of temperature, humidity, and rainfall patterns throughout seasons, years, or decades is referred to as climate. In other terms, climate is the long-term average of weather. Climate varies from area to region and from season to season. Climate influences water potential, natural vegetation, cropping patterns, land use, housing and other infrastructure, industrial location, racial features, and so on. Climate data is comprised of the mean value of climatic variables such as temperature, wind, and rainfall for various places and time periods.

11.4 Climatic Variables

11.4.1 Light

The visible region of the solar spectrum, with wavelengths ranging from 0.39 to 0.76, is what we refer to as light.

In order to maintain plants, light is a crucial component. The quantity of light a plant receives determines the rate of development and how long it stays active. Photosynthesis, a plant's most fundamental metabolic activity, uses light energy. The most significant pigment, chlorophyll, which absorbs solar energy and transforms it into the potential energy of carbohydrates, depends on it for its synthesis.

Four factors need to be taken into account when analyzing how light affects plant growth: intensity, duration, direction, and quality.

11.4.1.1 Light Intensity

A standard candle is used as a benchmark for measuring light intensity. "Meter candle or Lux" is the term used to describe the amount of light received at one meter from a normal candle. The device used to measure light intensity is the lux meter, and the amount of light at one foot from a normal candle is known as a "foot candle" or 10.764 luxes.

The production of plant food, stem length, leaf color, and blooming are all influenced by light intensity. Low light intensity typically results in spindly plants with light green foliage. Similar plants that are cultivated in intense light typically have bigger, dark green leaves, and shorter, better branches. Low light intensity slow down photosynthesis, which results in slower development. Likewise, extremely high intensity results in higher respiration. Additionally, it accelerates the rate at which plants lose water through transpiration. High intensity light has several negative effects, but oxidizing cell contents, often known as "solarization," is the most detrimental. This oxidation is referred to as photooxidation and is distinct from respiration.

11.4.1.2 Classification of Plants Based on Light Intensity

- *Sciophytes* (plants that prefer shade): These species do well in situations of partial shade. Buck wheat, betel vine, etc.
- *Hetrophytes* (sun-loving): Several types of plants generate their highest levels of dry matter under high light intensities. When the moisture is present at the ideal level, for example, rice, sorghum, and maize.

11.4.1.3 Quality

White light is split up into several color wavelengths as it passes through a prism. It is referred to as the visible portion of the solar spectrum. Violet, blue, green, yellow, orange, and red are some of the hues that make up the spectrum. The violet-blue and orange-red wavelengths are the main ones that are absorbed and utilized in photosynthesis.

The development of plants is negatively impacted by shortwaves like X and gamma rays as well as longer light waves like infrared. The best light for growth is red, followed by violet and blue. Many fungi and bacteria are killed by ultraviolet radiation and shorter wave lengths.

11.4.1.4 Day Length/Duration

The length of the day or the duration of light received by plants is also important. Some plants, such as Christmas cactus, only flower when the day is 12 h or less (short-day plants). Some plants only blossom when the days are longer than 12 h (long-day plants), but others are completely unaffected by day length (day-neutral plants). Photoperiodism is the reaction of plants to the duration of the day and night. The plants are categorized depending on their degree of responsiveness to day duration, as shown below.

- Long-day plants are those that develop and produce normally when the photoperiod exceeds the critical minimum (more than 12 h), for example, potato, sugar beet, wheat, barley, and so on.
- Short day plants: Plants that develop normally when the photoperiod is shorter than the critical maximum (less than 12 h). Rice, sorghum, cotton, and sunflower are examples of crops.
- Indeterminate/day neutral plants: Plants that are not impacted by photoperiod, for example, tomatoes.

11.4.2 Atmospheric Temperature

According to the definition of temperature, it is "The measure of speed per molecule of all the molecules of a body." Heat, on the other hand, is defined as "the energy arising from random motion of all the molecules of a body." It relates to the heat energy's intensity.

11.4.2.1   Role of Temperature in Plant Production

- Vegetation and agricultural plants are distributed according to temperature.
- Air temperature has an impact on blooming, leaf growth, and output.
- One of the crucial factors that affects all phases of crop growth, development, and reproduction is the surface air temperature.
- Air temperature affects crop biochemical processes, which can increase by two or more for every 10 °C increase.
- Temperature affects how quickly gases and liquids diffuse.
- Most crops have upper and lower temperature thresholds below or above, which they may not grow, as well as an ideal temperature at which crop growth is at its highest. These are referred to as cardinal temperatures, and various crops require various temperatures.
- The majority of crop plants are afflicted by pests and diseases under high temperatures and high humidity.
- The metabolism and respiration increase at high nighttime temperatures.
- Temperature affects a substance's ability to dissolve.
- Temperature affects how different systems and chemicals are in equilibrium.
- The stability of enzyme systems in plants is influenced by temperature

Thermoperiodism is the response of a living organism to regular fluctuations in temperature, whether day or night or seasonal (Table 11.1).

11.4.2.2   Diurnal and Seasonal Variation in Temperature

1. Temperature changes from day to night are known as diurnal variation and are a result of the Earth's daily rotation. The sun heats the Earth during the day (by solar radiation), yet it continues to lose heat due to terrestrial radiation. Solar and

Table 11.1 The cardinal temperature of some important crops (Mavi & Tupper, 2004)

Plant	Cardinal temperature (°C)		
	Minimum	Optimum	Maximum
Wheat	3–4.5	25	30–32
Barley	3-4.5	20	38–40
Maize	8–10	32–35	40–44
Rice	10–12	30–32	36–38
Tobacco	13–14	28	35
Sugar beet	4–5	25	28–30
Peas	1–2	30	35
Oats	3–4	25	30
Sorghum	8–10	32–35	40
Lentils	4–5	30	36
Carrot	4–5	8	25
Pumpkin	12	32–34	40

Source: Adapted from Bierhuizen (1973)

terrestrial radiation imbalances are what cause warming and cooling. The surface warms up during the day as solar radiation outweighs terrestrial radiation. Even though solar radiation is no longer present at night, terrestrial radiation still exists and keeps the surface cool. Around sunrise, the air temperature is at its lowest; from there, it rises steadily until it reaches its highest point. Though the maximum solar radiation is reached at noon, the highest air temperature is recorded between 1300 and 1400 h. After reaching its peak, there is a gradual decrease in temperature until dawn. As a result, the daily march shows one temperature at its highest and lowest.

> *Diurnal Temperature Range*: The difference between the daily maximum and lowest temperatures is known as the diurnal temperature range. The largest diurnal variation is seen in humid, desert, and city environments close to big bodies of water.
>
> *Mean Daily Temperature*: Average daily temperature during a period of 24 h.

2. Seasonal variation: In addition to its daily rotation, the Earth also completes one orbit of the sun every year. The axis of the Earth tilts toward the plane of orbit; thus, the hemispheres experience incoming solar radiation at different angles according to the seasons. Because it receives more solar energy than the Southern Hemisphere does, the Northern Hemisphere is warmer in the months of June, July, and August. The converse is true in December, January, and February, when the Southern Hemisphere is warmer and gets more solar radiation.

11.4.2.3   Variation of Air Temperature with Latitude

Horizontal air temperature distribution depends upon the latitude. Isotherms are the lines connecting the points of equal temperature. The angle of incoming solar radiation varies geographically due to the curvature of the Earth. The sun is more directly above in equatorial regions than at higher latitudes because the Earth is basically spherical. As a result, equatorial regions are the hottest and get the most radiant radiation. Higher latitudes receive less energy from the sun's slanted beams, with the poles receiving the least. So, from the warm Equator to the cold poles, temperature varies with latitude.

11.4.2.4   Vertical Temperature Distribution

Temperature normally falls as height increases. It is known as the typical lapse rate. The typical rate of decrease in temperature in the troposphere ascending to the tropopause is roughly 6 °C per kilometer. This is also known as a vertical temperature gradient. The usual lapse rate is not always the same; it varies based on height, season, latitude, and a variety of other local conditions.

11.4.2.5   Temperature Inversion

The conditions are sometimes reversed, and the regular lapse rate is inverted. It is known as temperature inversion. In other words, the temperature inversion is the situation in which the temperature abruptly rises rather than falls in the air.

11.4.2.6   Significance of Atmospheric Temperature

11.4.2.6.1   High Temperature

- Heat stress is frequently described as an increase in temperature that lasts for long enough to permanently harm plant growth and development. Heat shock or heat stress generally refers to a brief rise in temperature, often 10–15 °C over ambient. Worldwide agricultural output is seriously threatened by heat stress brought on by high ambient temperatures.
- High temperatures cause plants to dry out, disrupt physiological processes like photosynthesis and respiration, and accelerate respiration, which causes reserves of food to be depleted quickly.
- Sun clad: High daytime temperatures and low nighttime temperatures cause damage to the stem's bark.
- Stem griddle: Due to the hot soil, the stem at ground level scorches all over. By damaging conductive tissues, it kills plants. A prime instance is immature cotton seedlings, where this kind of damage is fairly prevalent when the soil temperature in sandy soil reaches 60 °C.
- Various plant diseases are brought on by high temperatures and high humidity.

11.4.2.6.2   Low Temperature

- Chilling injury: The injury caused in the plants due to the drop of the temperature below 15–20 °C but above the freezing point is known as plant chilling injury. When exposed to low temperatures for an extended period of time, plants that are adapted to hot climates are found to be killed, severely injured, or develop chlorotic condition (yellowing), such as chlorotic bands on the leaves of sugarcane, sorghum, and maize in winter months when the night temperature is below 20 °C. Chilling injury to fruits causes surface pitting, lesions, discoloration, susceptibility to decay organisms, and a reduction in storage life. Chilling injury causes fruits to ripen abnormally. Chilling injury occurs at temperatures ranging from 8 to 12 °C for tropical fruits such as banana, avocado, and mango and 0–4 °C for temperate zone fruits such as apple (Kozlowski, 1983).
- Freezing injury: It occurs at a temperature below the freezing point and is commonly observed in the plants of temperate region. Water in the intercellular spaces of plants freezes into ice crystals when they are exposed to very low temperatures. Cells die as a result of the protoplasm being dehydrated.

- During periods of freezing and thawing caused by snowfall and extremely low temperatures, plants uproot themselves.
- Suffocation: In colder climates, the soil surface often develops a thick layer of ice or snow throughout the winter. As a result, oxygen cannot enter, and the crop suffers from lack of oxygen. The diffusion of CO_2 outside the root zone is stopped when ice comes into contact with the root. The respiratory activity of roots ceases as a result of this, and it leads to buildup of harmful substances

11.4.3 Atmospheric Humidity

Humidity: The amount of water vapor present in the atmosphere is referred to as atmospheric moisture or humidity. There are several ways of describing air's humidity, and they are as follows.

1. *Absolute humidity*: Absolute humidity is the amount of water vapor in a given volume of moist air. It is measured in terms of grams of water vapor per cubic meter or cubic foot.
2. *Specific humidity*: Specific humidity is defined as the weight of water vapor per unit weight of wet air. The unit of measurement is grams of water vapor per kilogram of air (g/kg).
3. *Relative humidity*: It is the ratio of the water actually present in the air to the greatest quantity that might be present at the temperature of the air.
4. *Dew point °C*: The temperature at which the air must be cooled in order for the moisture present to represent saturation. In other words, it is the temperature at which the vapor pressure of water is the same as the atmospheric pressure. Dew is deposited on accessible surfaces when air cools down below the dew point, or else it produces cloud droplets (Linacre & Geerts, 2003).

11.4.3.1 Role of Humidity in Plant Production

- The crops that are cultivated in a certain area depend on the humidity.
- It affects a number of physiological processes in agricultural plants, such as transpiration.
- The presence of excessive humidity has negative impacts. It promotes the growth of several parasitic and saprophytic fungus, bacteria, and pests, whose expansion severely damages agricultural plants, for example, potato blight disease.
- A significant factor in determining potential evapotranspiration is the humidity. As a result, it establishes the water needs of crops.
- Because crop evapotranspiration losses are dependent on atmospheric humidity, crops require less water for irrigation when the humidity is high.
- It has an impact on the water potential of plants inside.

- Crops under moisture stress can survive for longer under high humidity conditions. However, extremely high or extremely low relative humidity does not promote higher crop yields.
- Moderate relative humidity of 40% is almost safe for all the crops.

11.4.3.2 Variations in Atmospheric Humidity

In the absence of severe weather changes, the variations in humidity occur throughout the day. The specific humidity and dew-point temperature stay basically constant from day to night. However, as the temperature drops at night, the saturation vapor pressure falls, causing relative humidity to rise. The highest humidity levels are frequently found before daybreak, when the temperature is at its lowest.

Seasonal fluctuations in temperature and seasonal changes in atmospheric circulation both influence seasonal differences in humidity. In areas with temperate climates, summer air is warmer and has a higher saturation vapor pressure than winter air. Since the average relative humidity levels in summer and winter rarely differ significantly, summer's specific humidity is typically much higher. When onshore winds transport moisture-laden air from the ocean to land areas during the monsoon season, relative and specific humidity tend to rise dramatically in monsoon climates.

11.4.4 Atmospheric Pressure

The weight of the air that is vertically above a unit area centered at a point is the atmospheric pressure. A pressure of 1.034 gm/cm^2 is applied to the ground by the weight of the air. The unit of measurement is millibar (mb), which is equivalent to 100 N/m^2 or 1000 dynes/cm^2. Different atmospheric pressures result from the sun's uneven heating of the planet and its atmosphere as well as the earth's rotation.

Isobars: "Isobars" are used to depict the pressure distribution on maps. Isobars are described as imaginary lines that are created on a map to connect locations with the same atmospheric pressure.

11.4.4.1 Role of Atmospheric Pressure in Meteorology

In meteorology, pressure is significant because its measurement reveals how much air is above. A decrease in pressure indicates a net loss from the entire air column above, which denotes an outgoing wind. The direction and force of the wind are indicated by differences in pressure at different locations. Surface pressures vary seasonally as well, and deviations from the typical yearly cycle, for example, indicate exceptional temperatures aloft that are important for weather forecasting.

11.4.5 Solar Radiation

Solar energy provides the light needed for plant physiological processes such seed germination, leaf expansion, stem and shoot growth, flowering, and fruiting as well as the temperature conditions needed for these processes.

As a regulator and controller of growth and development, solar radiation is crucial. The distribution of dry matter and the assimilation of nutrients are both influenced by solar radiation.

11.4.5.1 Role of Solar Radiation in Agriculture

1. It provides all of the energy required for the phenomena related to biomass production.
2. The real source of energy for the photosynthetic process is active radiations (PAR). Solar energy is efficiently converted into biomass by plants. The crop yield in a particular region is determined by radiation.
3. It additionally supplies the energy needed for the physical processes occurring in soil, plants, and the atmosphere.
4. It influences crop distribution on the earth's surface by modifying temperature distribution.

11.5 Condensation

Condensation is the conversion of water vapor into liquid. The opposite of vaporization, which occurs when liquid water transforms into a gas, is condensation.

11.5.1 Basic Process of Condensation

Condensation takes place as a result of loss of heat (latent heat of condensation). When humid air is cooled, it may reach a point where it can no longer store water vapor. The extra water vapor then condenses into liquid. It is referred to as sublimation if it immediately condenses into solid form. Condensation forms in free air as a result of cooling around extremely tiny particles known as hygroscopic condensation nuclei cool, for example, dust, smoke, pollen, and seawater salt fragments. Dust, smoke, pollen, and sea salt make ideal nuclei because of their capacity to absorb water. Condensation can also occur when moist air makes contact with a colder object and when the temperature is very close to the dew point. Therefore, the amount sof cooling and the relative humidity of the air have an impact on condensation. After condensation, water vapor or moisture in the atmosphere takes the forms of dew, frost, fog, and clouds.

11.5.2 Forms of Condensation

Condensation can occur when the dew point is both lower and higher than the freezing point, and this forms the basis for classification.

- When the temperature falls below the freezing point, white frost, snow, and certain clouds (cirrus clouds) form.
- When the temperature is greater than the freezing point, dew, fog, and clouds form.

11.5.2.1 Dew

Dew is a term used to describe moisture that is deposited as water droplets on cooled solid objects like rocks, grass, and plant leaves. Water changes from a vapor to a liquid to generate dew. As the temperature drops and things cool down, dew develops. The air around an item cools down when it becomes sufficiently cold. In comparison to warmer air, colder air has a worse capacity to hold water vapor. As a result, water vapor in the air surrounding cooled objects condenses.

11.5.2.2 Frost

Frost is formed when the dew point is below freezing point. The moisture condenses to form small ice crystals known as Frost. The excess moisture in the air condenses directly to form small ice crystals in this process. Standing crops, like potatoes and others, are severely harmed by this kind of condensation.

11.5.2.3 Fog

Condensation occurs on fine dust particles when the temperature of an air mass that contains a significant amount of water vapor suddenly drops.

In other words, the fog is a cloud whose base is at or extremely close to the ground. Visibility drops to under a kilometer.

11.5.2.4 Mist

When condensation happens close to the surface of the earth, it produces mist, which is composed of tiny water droplets suspended in the air. In the mist, visibility is more than one km but less than two kilometers.

11.5.2.5 Smog

Smog is created when fog is coupled with smoke, dust, carbon monoxide, sulfur dioxide, and other contaminants. Large cities and industrial regions frequently experience smog. The respiratory system is impacted.

11.5.2.6 Clouds

A cloud is described as "a visible aggregate of tiny droplets of water, or particles of ice, or a mixture of both floating in the free air." Ice crystals, some of which may be as long as a tenth of a millimeter, make up the clouds that are found in higher and very cold regions of the sky.

11.5.2.6.1 Formation of Clouds

In some form or another, water is always present in the Earth's atmosphere. Water molecules, on the other hand, are too tiny to create the bonds necessary for cloud droplets to form on their own. They require a "flatter" surface on which to condense, ideally one with a radius of at least one micrometer (one millionth of a meter). They are known as cloud condensation nuclei.

Small solid and liquid particles, such as smoke from fires or volcanoes, ocean spray, or microscopic flecks of wind-blown soil, make up the vast majority of these nuclei. They are around 1/100th the size of a cloud droplet where water condenses and are hygroscopic, which means they attract water molecules.

As a result, each cloud drop has a tiny particle of dirt, dust, or salt crystal at its center. The main component of a cloud droplet is still pure water, even with the condensation nuclei present.

11.5.2.6.1.1 Function of Temperature

A cloud cannot form just because water-attractive nuclei are present; the air temperature must also be below the dew point, the saturation point at which evaporation equals condensation.

There are many ways for air to get saturated. The most typical method involves surface-to-atmosphere air rising and cooling.

The lower pressure higher in the atmosphere causes a block of air (a "parcel") to expand as it rises. It cools as a result of its expansion. The adiabatic process occurs when the temperature of a parcel varies owing to expansion or contraction rather than heat transmission between the parcel and the surrounding atmosphere.

The parcel's air temperature will drop by 5.5 °F (9.8 °C per kilometer) for every 1000 feet of elevation gain until it hits saturation. The term "dry lapse rate" refers to this.

Water vapor will condense onto the cloud condensation nuclei once the parcel achieves saturation temperature (100% relative humidity), which will lead to the creation of a cloud droplet.

The atmosphere, however, is always in motion. As air rises, drier air is incorporated or entrained into the ascending parcel, resulting in continuous condensation and evaporation. Cloud droplets are continually developing and dispersing as a result. Clouds form and expand when more water condenses on nuclei than evaporates from them. Clouds, on the other hand, vanish if there is more evaporation than condensation. This explains why clouds are continually changing in appearance and disappearing.

11.5.2.6.2   Classification of Clouds

The World Meteorological Organization (WMO) divided the clouds into ten groups based on their height and appearance. According to their height, clouds may be divided into the following four primary groups: family A, B, C, and D. Within each of these basic categories, there are other subcategories.

- High level clouds (altitudes of 5–13 km) are one of the four cloud groups that are located at various tropospheric altitudes.
- Medium level clouds (2–7 km)
- Low level clouds (0–2 km)
- Clouds with long vertical extensions (0–13 km)

11.5.2.6.2.1   WMO Cloud Classification (1957)

- *Family A*: High clouds make up this group of clouds. In the tropics and subtropics, the mean lower level is 7 km, while the mean upper level is 12 km. There are three subgroups in this family:

  1. Cirrus (Ci): Ice crystals can be found in this kind of cloud. It has a delicate, silky, white fibrous, and feathery texture that gives it the illusion of being wispy and feathery. These clouds do not block the sun's beams from view. There is no precipitation.
  2. Cirrocumulus (Cc): These clouds also include ice crystals, similar to cirrus clouds. It resembles ocean waves or ruffled sand. It is translucent, has no shade, and has white spherical masses. The sky is like a mackerel.
  3. Cirrostratus (Cs): These clouds include ice crystals, just like the previous two. The whole sky is covered in what appears to be a milky white veil. It results in "Halo".

- *Family B:* Middle clouds make up this group of clouds. In the tropics and subtropics, the mean lower level is 2.5 km, while the mean upper level is 7 km. There are two subcategories in this family.

1. Altocumulus (Ac): Ice water can be found in these clouds. It features globular masses that are greyish or blue. It is also known as flock clouds or wool-packed clouds and has the appearance of a sheep's back.
2. Altostratus (As): These clouds have different concentrations of ice and water. It resembles a fibrous sheet or veil that is either grey or blue in hue. Shadows and coronas are produced. High and intermediate latitudes are where rain falls.

- *Family C*: Lower clouds make up this group. In the tropics and subtropics, these clouds reach a height of 2.5 km above the earth. There are three subcategories in this family.

   1. Stratocumulus (Sc): Water makes up these clouds. It seems softer and grayer than altocumulus, with big globular aggregates. Broken masses or long parallel rolls pressed together. While the air is calm above these clouds, there are powerful updrafts below.
   2. Stratus (St): Water also makes up these clouds. These clouds (clouds close to the ground) appear to be like a grayish white sheet covering the entire area of sky. It mostly appears in the winter, and there may be some drizzle.
   3. Nimbostratus (Ns): These clouds consist of ice or water crystals. Effectively reducing the amount of daylight, it appears as a thick, homogeneous veil of darkness and gray. The precipitation is constant. Occasionally, it seems to be an amorphous, fragmented piece of paper.

- *Family D*: Convection, or vertical development, is the cause of these clouds, which belong to the Family D. Assuming a 16 km upper level, the mean low level is 0.5 km. Two subcategories fall under this family.

   1. Cumulus (Cu): These clouds are made of water and have a flat foundation and a majestic white look. Uneven dome with a wool pack and a shadowed, black appearance below. It resembles a cauliflower. Usually, these clouds grow to become cumulonimbus clouds with flat bases.
   2. Cumulonimbus (Cb): These clouds have water at their lower levels and ice at their highest levels. These clouds grow vertically and have thunderheads with imposing anvil tops. During the summer, these clouds create ferocious winds, thunderstorms, hail, and lightning (Fig. 11.4).

11.6   Precipitation

Continuous condensation in open air aids in the condensed particles' size expansion. They descend to the earth's surface when the air's resistance to gravity is insufficient to keep them suspended. Precipitation is the discharge of moisture after the condensation of water vapor. This can happen in either liquid or solid form.

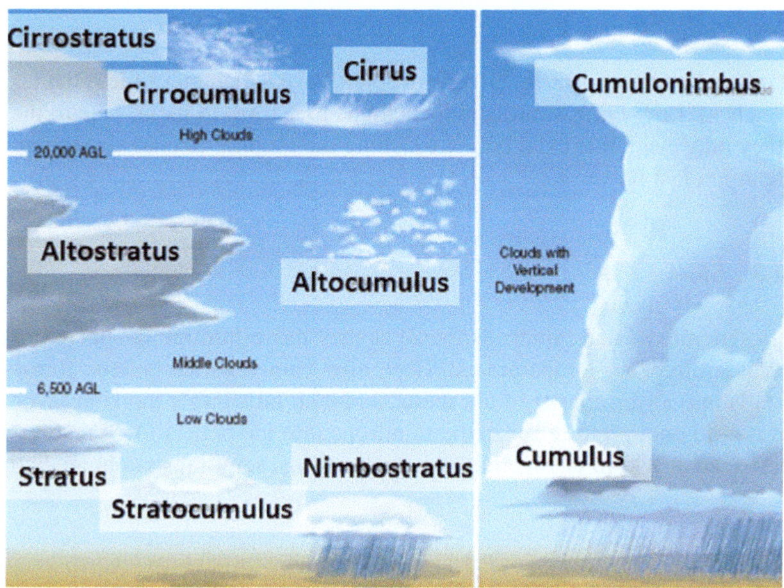

Fig. 11.4 Various classification of clouds based on their elevation and appearance. (Goswami, 2021)

11.6.1 Types of Precipitation

11.6.1.1 Rain

Rain is any liquid that falls from the sky's clouds. Rain is defined as water droplets of 0.5 mm or greater. Rainfall is precipitation in the form of droplets of water. Rainfall may be continuous or showery. Showers are more prevalent over land in the spring and summer. They can be categorized as "isolated" if they only impact ten percent or less of the region, "scattered" if they happen in between 10% and 50% of the area, or "widespread" (Linacre & Geerts, 2003).

11.6.1.2 Snow

Ice crystals in the shape of flaky particles, with an average density of 0.1 g/cc, make up snow. In colder regions and at higher elevations, it is a significant type of precipitation that develops. Typically, cirrus clouds that are high, thin, and feeble are seen with snow.

11.6.1.2.1 Drizzle

Light rain is known as drizzle. Drizzle droplets have a diameter of less than 0.5 mm (0.02 inches). Low-lying stratocumulus clouds give birth to them. Due to their tiny size, they may evaporate even before touching the ground. In cold air temperatures, drizzle might last for days.

11.6.1.3 Sleet

Sleet is generated when raindrops freeze as they move through the air in the atmosphere at subfreezing temperatures. Sleet, also known as ice pellets, forms when snow falls into a warm layer of air, melts, and then falls into a freezing layer of air, which is cold enough to refreeze the raindrops into ice pellets. As a result, sleet is characterized as a kind of precipitation made of tiny, semitransparent ice balls.

11.6.1.4 Hail

Ice, in one of three forms, is hail. There may be three types of hail: (i) true hail, which is bigger, opaque, and hard; (ii) soft hail, or graupel, which consists of tiny, slushy, frozen cloud droplets; and (iii) pellets of frozen rain up to 6 mm in diameter, which are basically huge sleet (Linacre & Geerts, 2003). In the winter or during cold weather, hailstones are more common. They seriously harm crops by pulling leaves apart and lowering the value of those leaves.

11.7 Weather Aberrations

A weather aberration is an event of exceptionally severe weather or climatic conditions that can have catastrophic effects on populations, agricultural systems, and natural ecosystems. Weather aberration includes unexpected, unusual, extreme, or unseasonal weather.

11.7.1 Flood

Rivers that overrun their banks cause floods.

This occurs when there is an unusual amount of runoff into the river from a large catchment area as a result of severe rainfall or the quickly melting of snow upstream. "Flash floods" are described as abrupt, transient river rises. Globally, floods result in a large number of fatalities, and the number rises as more people move into flood-prone areas.

Where the earth is too impermeable or already too saturated to absorb extra precipitation, floods are more likely to occur.

11.7.2 Drought

Drought is generally understood to be a complicated phenomenon that happens when there is a protracted lack of precipitation and a high rate of evaporation. This disrupts the water supply to plants by causing anomalous water loss from water bodies, decreasing the water table, and dehydrating the soil's root zone.

There are several definitions for droughts.

1. When there is less rain than average in terms of the degree of variability at the place, this is referred to as a meteorological drought. Every state experiences a particular amount of typical rainfall in each region. Planning the region's or area's cropping pattern will be based on this. Meteorological drought occurs when yearly precipitation falls considerably short of a threshold (75% of the climatologically predicted normal rainfall) across a large region.
2. Soil moisture levels at crucial periods of crop growth define an agricultural drought. As a result, it is affected by past runoff and evaporation (i.e., radiation, temperature, humidity, and wind) as well as rainfall. Agricultural drought is caused by insufficient rainfall and is followed by a soil moisture deficit. As a result, soil moisture falls short of meeting crop demands during the growing season.

   The drying up of streams, rivers, reservoirs, lakes, and wells is referred to as hydrological drought (Linacre & Geerts, 2003).

11.8   Indian Monsoon

11.8.1   Introduction

Large amounts of seasonal rainfall and a change in wind direction are two classic characteristics of the monsoon climate. Most importantly, this rain has been a major source of fresh water for many human endeavors, including agriculture. The Indian Subcontinent is at the center of the Southeast Asian summer monsoon system, and the monsoon trough stretches from northern India to Indochina to the Western Tropical Pacific (WTP). The summer monsoon season, which runs from June to August 1, is when a significant portion of the yearly rainfall falls, with two separate maxima. One is situated above the Bay of Bengal, with rain falling into eastern and central India to the northwest, while the other is along the west coast of India where a lower level moist wind meets the Western Ghat Mountains (Saha & Bavardeckar, 1976). Rainfall in the rest of the Indian subcontinent is significantly less. Technically, the name "monsoon" came from the Arabic word "Mawsim," which was used to describe seasons. The term "monsoon" refers to the "seasonal winds" that produce rain in different places of the world during particular times of the year. In India, the monsoon occurs by northeast monsoon winds in the winter and southwest monsoon winds in the summer.

11.8.2 Monsoons in India

India's farmers eagerly await the arrival of the monsoon season. Being primarily an agricultural nation, India looks forward to a good monsoon. The Indian monsoon has two types:

- Northeast monsoons
- Southwest monsoons

11.8.2.1 Southwest Monsoons

The majority of India's rainfall is heavy, and this is mostly due to the southwest monsoon winds. Extreme summer heat on the Tibetan Plateau is what triggers the Southwest monsoon. As a result, low pressure and persistent high pressure develop in the Indian Ocean's southern direction. The Southwest monsoon's rains are brought on by strong air currents that move through the warm equatorial ocean, which causes evaporation. The southwestern winds split into two after passing across the equator and travel through the Arabian and Bay of Bengal seas. This occurs as a result of the Western Ghats being struck by southwest monsoon winds. High amounts of evaporation from the ocean's surface produce water vapor, and the southwest monsoon winds, which are then laden with water vapor, gradually begin cooling while rising above the land surface and travelling northward. When the air reaches saturation and is unable to retain moisture, heavy rains fall.

- At times, there are numerous heavy showers of rain, resulting in flooding in the area. During the Southwest monsoon, Kerala is the first Indian state to get rain. The southwest monsoon is forecast to start around the beginning of June and last until the end of September.
- During the Southwest monsoon, the Tamil Nadu coast normally remains dry since Tamil Nadu receives its rainfall.

11.8.2.2 Northwest Monsoons

During the winter season, high pressure builds up over the Siberian and Tibetan Plateaus, causing the Northeast monsoon to form. Rains are brought to the Southeast coast of India by the Northeastern winds. The Southeast coast includes parts of the Tamil Nadu coast and the south shore of Seemandhra. During the northeast monsoon, winds sweep from sea to land, carrying the moisture generated by the Indian Ocean. From October through December, the northeast monsoon mostly impacts South India, producing rainfall in Puducherry, Karaikal, Tamil Nadu, Andhra Pradesh, Yanam, Mahe, and Karnataka. The three months indicated above result in high rainfall in the Tamil Nadu area, accounting for 48% of total rainfall in a year (Fig. 11.5).

11.8.3 Importance of Monsoon

India is a country with an agricultural economy; thus, rainfall is quite important. There is no denying the monsoon season's significance in India.

- India eagerly awaits the monsoon since it has a significant impact on the development of our essential crops and vegetables. Comparatively to any other country in the world, India uses a sizable portion of its territory for agricultural operations.
- Since a large portion of Indians work in agriculture, a poor monsoon reduces income and has an impact on the life of many people.
- Less rainfall causes crops to be wasted unnecessarily.
- India's hydroelectric power generation, which makes up at least 25% of the country's electrical production, can increase during a strong monsoon.

11.9 Weather Forecasting

The most significant element affecting agricultural operations and crop yield is the weather. A sizable amount of the harvest is lost because to unusual weather. Depending on the crop, preharvest losses might be between 10% and 100%. Rains and an excessive amount of humidity are the major causes of postharvest loss.

A weather forecast is a forecast of the weather for the upcoming several days. Even though losses brought on by weather factors cannot be completely avoided, they can be reduced by adjusting for the weather using timely and accurate weather forecasts. The weather predictions also offer recommendations for long-term seasonal planning and crop selection that is compatible with the predicted climatic conditions.

11.9.1 Importance of Weather Forecasting

A place's weather has significant social and economic effects. The weather is one of several variables that affect crop yield, and its irregularities alone can account for up to 50% of variations (Chandrasekaran et al., 2018). Unseasonably high

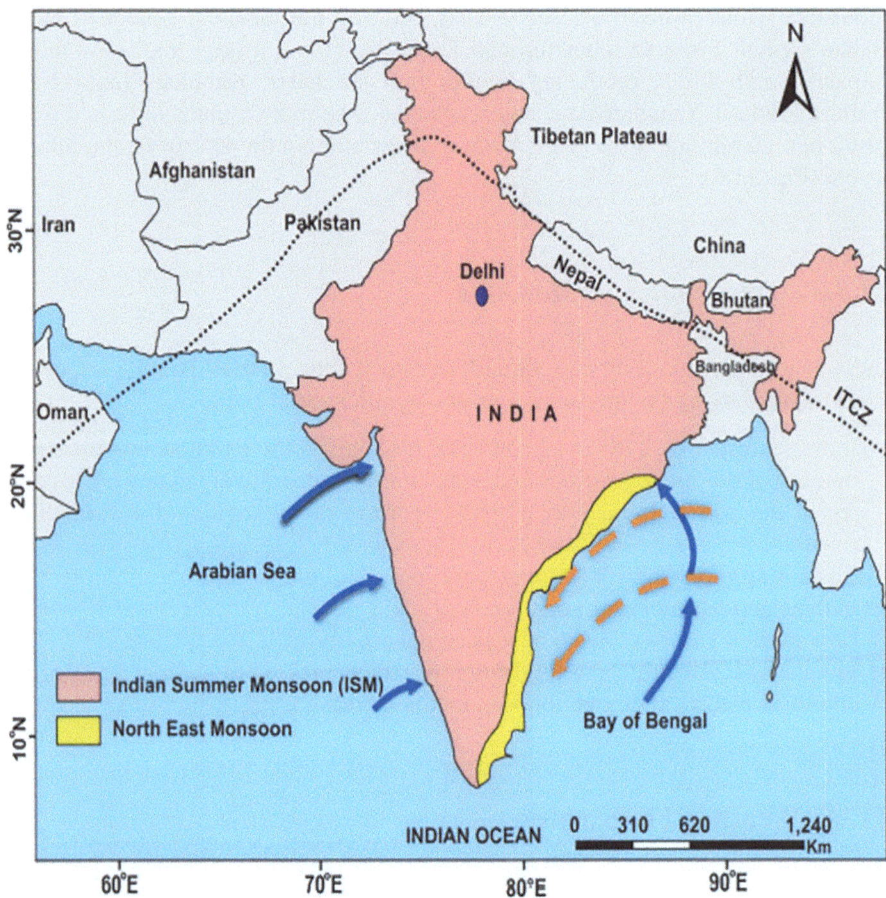

Fig. 11.5 Behavior of monsoon over India. (Singh et al., 2018)

temperatures may cause plants to produce less and more pests on farms. Farmers can decide what to grow and when to plant it with the aid of rainfall and temperature forecasts. Farmers can appropriately schedule agricultural activities like planting, irrigation, fertilizer application, pruning/weeding, harvesting, and animal mating by using weather forecasts. Following application highlight the importance of weather forecasting:

1. *Relationship between fertilizer timing and weather forecasting*: Weather forecasting aids farmers by letting them know when, how much, and what kind of fertilizer to use at what time. The development of the crop may be hampered by fertilizer application and scheduling errors. The field has to be damp enough for the fertilizer to penetrate the soil while being dry enough to prevent the fertilizer from washing away. Accurate weather information may help farmers choose the best times to conduct their regular business.

2. *Pest control*: The ability to predict the weather is helpful in preventing pests and other crop diseases from spreading over a field. Crop-destroying pests can be influenced by weather conditions. When to use pesticides can be determined with the use of this information.
3. *Field workability*: Field workability is the availability of days that are good for conducting fieldwork. The two most crucial variables are temperature and soil moisture. Accurate field-level meteorological data can help farmers assess the viability of their farms and streamline daily operations.
4. *Scheduling of irrigation*: A method of artificially applying water to land for farming and agricultural production is called irrigation. Variability in the weather has an impact on irrigation needs and agricultural production. Climate change is a factor to which all farmers must adapt. Drought, which lasts for extended periods of time, has a significant influence on the irrigation system. Therefore, the likelihood of losses is much lower than anticipated if their proper forecast is done.

11.9.2   Types of Weather Forecasting

There are three forms of agricultural weather forecasting.

1. *Short-range forecast*: It has an accuracy of 70–80% and is valid for 24–48 h. The temperature, wind speed and direction, sunlight length, timing and quantity of precipitation, and relative humidity are all highlighted in the short-range prediction. It provides irrigation planning, timing adjustments for agricultural tasks, and frost protection for plants.
2. *Extended forecast*: It has a 5-day validity window and a 60–70% accuracy rate. The focus is placed on the kind of weather, typical weather, the order of rainy days, and farming dangers such high winds and prolonged dry or wet spells. This kind of prediction aids in making decisions regarding irrigation planning, harvesting, and spraying timing, as well as sowing time and depth.
3. *Long-range forecast*: It is applicable beyond 10 days up to a month and season. It focuses on temperature and precipitation anomalies. This prediction will be useful in determining soil moisture management, irrigation scheduling, crop selection, irrigation management with restricted water supply, and cropping pattern and crop output (Table 11.2).

MCQs

1. The temperature drops as we ascend higher into the troposphere at the rate of

   - 6.5 °C/1000 m
   - 10 °C/1000 m
   - 0.65 °C/1000 m
   - 8 °C/1000 m

Table 11.2 Types of forecasting (Chandrasekaran et al., 2018)

Types of forecast	Validity period	Main users	Predictions
1. Short range (a) Now casting (b) Very short range	Up to 72 h 0 to 2 h 0 to 12 h	Farmers, marine agencies, general public	Rainfall distribution, heavy rainfall, heat and cold wave, thunder storm.
2. Medium range	Beyond 3 days and up to 10 days	Farmers	Occurrence of rainfall, temperature.
3. Long range	Beyond 10 days and up to a month and season	Planners	This forecasting is provided for Indian monsoon rainfall. The outlooks are usually expressed in the form of expected deviation from normal condition

2. The coldest spot on Earth, with an average temperature of roughly −85 °C is

   - Thermosphere
   - Mesopause
   - Exobase
   - Mesosphere

3. Which of the following is a shade loving plant?

   - Sorghum
   - Maize
   - Rice
   - Buck wheat

4. Which of the following is a high cloud?

   - Cumulus
   - Altocumulus
   - Cirrus
   - Stratus

5. The medium range weather forecasting is done for

   - 72 h
   - 3–10 days
   - 20–25 days
   - Up to 3 days

6. During the Southwest monsoon, ______ is the first Indian state to get rain

   - Tamil Nadu
   - Kerala
   - Gujrat
   - Andhra Pradesh

7. Which of the following clouds are associated with ferocious winds, thunderstorms, hail, and lightning during summer.

- Strato cumulus
- Cumulus
- Nimbostratus
- Cumulonimbus

8. Frost is formed when the dew point is ______ freezing point.

- Equal to
- Below
- Above
- None of the above

9. Isobar is________?

- Line of equal pressure
- Line of equal rainfall
- Line of equal temperature
- None of the above

10. The word Monsoon has a

- Indian origin
- Latin American origin
- Arabic
- Indo-Malayan origin

References

Ahmad, L., Kanth, R. H., Parvaze, S., & Mahdi, S. S.(2017). *Experimental agrometeorology: A practical manual.* 1, 2, 89. - Springer Cham.

Bierhuizen, J.F. (1973). *Environmental effects on crop physiology.* Academic Press.

Boyer, M., Atkinson, K. & Arumainayagam, C. (2019). *The role of low-energy (< 20 eV) electrons in atmospheric processes.* CRC Press/Jenny Stanford Publishing.

Chandrasekaran, B., Annadurai, K., & Somasundaram, E. (2018). *A textbook of agronomy,* 200, 205, 210, 223 - New Age International Private Limited

Fullarton, L. (2016). Legislating climate change: Australia's renewable energy target legislation examined by a solar farmer. In *23rd Sustainable Economic Growth for Regional Australia (SEGRA) Conference.*

Goswami, P. (2021). A survey of modeling, rendering and animation of clouds in computer graphics. *The Visual Computer, 37,* 1931.

https://www.noaa.gov/jetstream/clouds/how-clouds-form

Hyman, A. (2017). *The climatology handbook.* 1–3. Callisto Reference.

Kozlowski, T. T. (1983). Reduction in yield of forest and fruit trees by water and temperature stress. In C. D. Raper & P. J. Kramer (Eds.), *Proceedings of the workshop on crop reactions to water and temperature stresses in humid and temperate climates* (pp. 50–67).

Linacre, E., & Geerts, B. (2003). *Climates and weather explained.* 111, 189, 192, 208, 209 - Routledge.

Matsuoka, D. (2017). Extraction, classification and visualization of 3- dimensional clouds simulated by cloud-resolving atmospheric model. *International Journal of Modeling, Simulation, and Scientific Computing, 8*, 1750051:1.

Mavi, H. S., & Tupper, G. J. (2004). *Agrometeorology – Principles and applications of climate studies in agriculture* (1st ed., pp. 47–54). The Haworth Press.

Molga, M. E. (1962). *Agricultural meteorology*. Pergamon Press.

Saha, K. K., & Bavadeckar, B. S. (1976). *Principles of agricultural meteorology*. Indian Council of Agricultural. Research.

Singh, A. K., Tripathi, J., Kotlia, B., Singh, K. K., & Kumar, A. (2018). Monitoring groundwater fluctuations over India during Indian Summer Monsoon (ISM) and Northeast monsoon using GRACE satellite: Impact on agriculture. *Quaternary International, 507*, 342–351. https://doi.org/10.1016/j.quaint.2018.10.036

Zaidi, N. W., Dar, M. H., Sudhanshu, S., & Singh, U. S. (2014). Trichoderma species as abiotic stress relievers in plants (Chapter 38). In V. K. Gupta et al. (Eds.), *Biotechnology and biology of trichoderma* (pp. 515–525).

Chapter 12
Climate Change and Agriculture

Abstract Climate change has become one of the major concerns of man in recent decades all over the world. The floods, drought, and desertification seriously threaten the livelihood of farmers and our food supply as well. As the climate continues to change, farm communities around the world will be increasingly challenged. Halfway through the chapter we get to know about the impacts of climate change like greenhouse effect, global warming, nonavailability of water, ozone depletion, etc. Toward the end, the chapter provides apt ways to mitigate climate change and how one can canvass for it, ultimately reversing it.

Keywords Climate change · Climate variability · Greenhouse · Ozone · Carbon dioxide · Global warming

12.1 What Is Climate Change?

Climate change refers to long-term changes in temperature and weather patterns. In this context, long term is a decade or longer. It is a slow and gradual process. Large-scale volcanic eruptions or shifts in the sun's activity can both naturally cause such changes. However, since the 1800s, human activities—primarily the use of fossil fuels like coal, oil, and gas—have been the main driver of climate change (UN-Climate Action).

According to the UNFCCC's official definition, climate change is any change that can be directly or indirectly attributed to human activity that modifies the global atmosphere's composition and that is in addition to natural climate variability seen over comparable time periods.

IPCC explains climate change as a change in the climate's state that can be detected by changes in the mean and/or variability of its parameters and that persists for a prolonged period, generally decades or longer.

The life of billions of people who rely on land for the majority of their requirements is significantly threatened by the frequent floods and droughts. Extreme disasters including cold and heat waves, landslides, forest fires, droughts, and floods

L. Ahmad et al., *Fundamentals and Applications of Crop and Climate Science*,
https://doi.org/10.1007/978-3-031-61459-0_12

regularly have a negative impact on the world economy. Increased atmospheric pollutants caused by greenhouse gas emissions such as carbon dioxide, chlorofluorocarbons (CFCs), hydrofluorocarbons (HFCs), ozone depletion and UV-B filtered radiation, volcanic eruptions, and indiscriminate deforestation are all contributing factors to weather extremes. Burning of fossil fuels releases greenhouse gases, which act like a blanket and trap the solar radiations causing temperature to rise. Global warming is simply the heating of the surface atmosphere caused by greenhouse gas emissions, which raises global atmospheric temperature over time. Such variations in surface air temperature and their long-term detrimental effects on rainfall is called as climate change.

12.2   Climate Variability

Climate variability refers to how different components of the climate (such as temperature and precipitation) vary from the average. Climate variability occurs as a result of natural and sometimes periodic variations in air and ocean circulation, volcanic eruptions, and other reasons. The phrase "climate variability" is frequently used to describe deviations in climatic data during a certain time period (e.g., a month, season, or year) when compared to long-term statistics for the same calendar period. These variations, which are commonly referred to as anomalies, are used to measure climate variability.

12.3   Comparison Between Climate Change and Climate Variability

Climate variability is concerned with changes that occur over shorter timescales, such as a month, a season, or a year. Climate change, on the other hand, evaluates changes that occur over a longer period of time, generally spanning decades or longer. In other words, variation is noise, whereas change is signal. (Year-to-year fluctuations are referred to as noise, whereas the trend is the signal.) (Fig. 12.1)

12.4   Causes of Climate Change and Climate Variability

12.4.1   Natural Causes

Solar Output: Changes in solar output are the most well-known external driver of climate change and variability. Solar output increased by 0.3% when compared to data from 1650 to 1700 AD (Chandrasekaran et al., 2018).

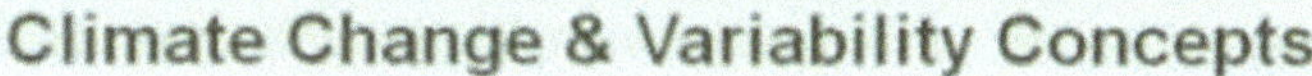

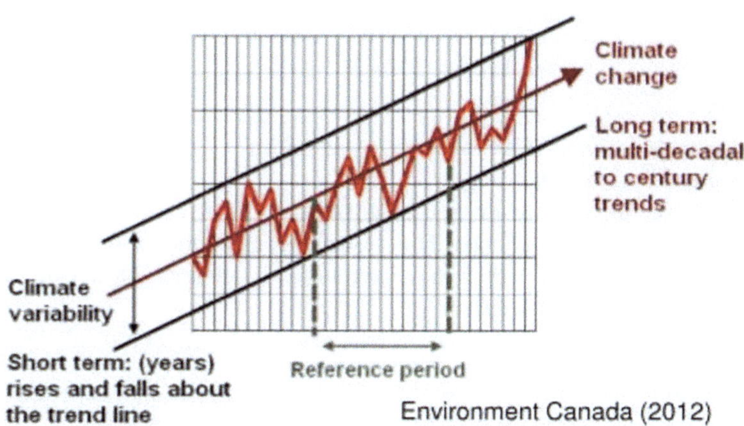

Fig. 12.1 Illustration of how climate change and climate variability occur together. (Mostafazadeh, 2018)

Changes in the Earth's Orbit and Rotation: The previous climate has been significantly impacted by modifications to the earth's orbit and rotational axis. For instance, past cycles of ice ages, in which the earth experienced long periods of cold temperatures (ice ages), as well as shorter interglacial periods (periods between ice ages) of relatively warmer temperatures, appear to have been primarily caused by changes in the amount of summer sunshine on the Northern Hemisphere, which are affected by changes in the planet's orbit.

Volcanic Activity: The climate has been significantly influenced by volcanoes, and in the distant past, massive amounts of carbon dioxide were emitted during volcanic eruptions. Particles from some violent volcanic eruptions, such SO_2, can be sent into the high atmosphere, where they can reflect enough sunlight back into space to keep the planet's surface cold for several years.

12.4.2 Human Causes

Greenhouse Gases: A portion of the outgoing infrared light that the earth and atmosphere release is trapped in the atmosphere by naturally occurring greenhouse gases. The main greenhouse gas is water vapor. Nitrous oxides (N_2O), ozone (O_3), methane (CH_4), and carbon dioxide (CO_2) are the other gases. The earth's surface and troposphere remain warmer than they would otherwise be because to these gases and clouds. The effectiveness with which the planet cools to space depends on changes in the amounts of certain greenhouse gases.

When greenhouse gas concentrations rise, the atmosphere absorbs more of the radiated energy from the surface of the earth. The quantities and distributions of greenhouse gases and aerosols in the atmosphere are changing as a result of human activities, particularly the burning of fossil fuels and deforestation. Human factors are to blame for a rise in global mean surface temperatures of 0.6 °C since the late nineteenth century (Mavi & Tupper, 2004).

Causes for Rising Emission

- Burning of fossil fuels: Carbon dioxide and nitrous oxide are the by-products of burning coal, oil, and gas.
- Forest destruction (deforestation): By absorbing CO_2 from the atmosphere, trees assist in controlling the climate. This advantageous impact is lost when trees are chopped down, and the carbon they have stored is released into the atmosphere, increasing the greenhouse effect.
- Rising livestock production: When cows and sheep digest their meal, they release a lot of methane.
- Nitrous oxide emissions are produced by nitrogen-containing fertilizers.

12.5   Impacts of Climate Change

It is anticipated that the earth's temperature would rise as CO_2 and other greenhouse gases become more concentrated. Crop production is dependent on the weather, and any change will have a significant impact on crop yield. The biological processes including respiration, photosynthesis, plant growth, reproduction, and water consumption are all impacted by elevated CO_2 and temperature. Depending on the latitude, CO_2 may have a positive influence or react in a different way. The following section discusses the impact of climate change including greenhouse effect, effect on water resources, agriculture, food, etc.

12.5.1   Greenhouse Effect

J.B. Fourler developed the "greenhouse effects" theory more than a century ago. Tyndall's research on the heat absorption by gases backed it up. The greenhouse effect happens when heat radiating from a planet's surface is partially trapped by greenhouse gases in the planet's atmosphere. Similar to the glass of a greenhouse, greenhouse gases absorb solar heat that is radiated from the surface of the Earth, trap it in the atmosphere, and keep it from escaping into space. The greenhouse effect maintains the Earth's temperature higher than it otherwise would be, allowing for the existence of life. Although many greenhouse gases are present in the atmosphere naturally, human activity helps them build up. As a result, the atmosphere's greenhouse effect is increased, which changes the climate of our planet and causes changes in snowfall and rainfall patterns, an increase in average temperature, and more extreme climate events like heatwaves and floods. Increased human activity results in higher levels of carbon dioxide, methane, nitrous oxide, chlorofluorocarbons (CFC), and other gases that raise the temperature and contribute to sea level rise. These gases, which are present in trace amounts, harm the environment by contributing to the greenhouse effect, stratospheric ozone loss, acid rain, smog, and corrosion.

12.5.1.1   Impacts of Green House Effect

- By altering the amount of polar ice caps, wind, rain, clouds, and ocean currents, the greenhouse effect will change the climate. These developments may have a significant influence on the entire world.
- The first concern, the loss of the ozone layer, puts the people of Earth in danger since it will lead to some significant issues in the future due to the introduction of damaging UV radiation.
- The second problem is global warming, which will cause the polar ice caps to melt in the future, endangering our coastal regions.
- Most of the world's water bodies are now acidic due to an increase in greenhouse gases in the atmosphere. Acid rain is produced when the greenhouse gases combine with precipitation. Water bodies get more acidic as a result of this.

The causes of exponential rise in greenhouse gases emission have been discussed in Sect. 12.4.2.

12.5.2   Ozone Depletion

A small layer of the atmosphere called ozone shields us from the sun. Since the dawn of time, the ozone layer has protected humans from the sun's most harmful UV radiation. Even now, it still does so. However, there has been an issue in recent decades. With increased use of items that release chlorofluorocarbons, or CFCs, the ozone layer is being destroyed. Currently, only a small portion of Antarctica is really covered by ozone due to the enormous hole that is developing over the continent. Even over countries with higher densities of population, including North America and Australia, the ozone layer is diminishing.

12.5.2.1   Causes of Ozone Depletion

- Chlorofluorocarbons (CFCs) and other ozone-depleting substances (ODS) released into the atmosphere contribute to ozone depletion. Refrigerants, insulating foams, and solvents are a few typical ODS. The sun's UV rays cause CFCs to disintegrate when they enter the stratosphere, releasing chlorine atoms that then interact chemically with ozone to begin cycles of ozone degradation that erode the ozone layer.

- It has been discovered that some natural processes, such as sunspots and stratospheric winds, degrade the ozone layer. The ozone layer is being destroyed due to volcanic eruptions as well.
- Methyl bromide (a pesticide), halons (found in fire extinguishers), and methyl chloroform (a solvent used in industrial operations) are further compounds that harm the ozone layer. On breaking down, methyl bromide and halons release bromine, which is 40 times more destructive to ozone then chlorine (Chandrasekaran et al., 2018).

12.5.2.2   Effects of Ozone Depletion

- Plants may be unable to grow, bloom, or perform photosynthesis due to intense UV radiation.
- As the ozone layer diminishes, individuals will be straightforwardly presented to the sun's perilous UV radiation. As a result, skin conditions, cancer, sunburns, cataracts, rapid aging, and weakened immune systems are potential health issues for humans.
- • Skin and eye cancer occur in animals that are directly exposed to UV light.

12.5.3   Global Warming

A portion of the solar radiation that strikes the Earth is absorbed by it, and some of it is reflected back into the atmosphere. These gases, which include ozone, methane, carbon dioxide, water vapor, and chlorofluorocarbons, are referred to as greenhouse gases because they absorb some heat and prevent it from leaving our atmosphere. These gases contribute to the atmosphere's heating, which causes global warming. Global warming is the gradual increase in the planet's surface temperature. Although this warming trend has been around for a while, human activity over the past century has caused it to pick up speed dramatically.

The average global temperature of Earth has reportedly risen by roughly 1 °C (1.8 °F) as a result of human activity since the pre-industrial era, and it is now rising by more than 0.2 °C (0.36 °C) every decade. Without a doubt, human activity since the 1950s has contributed to the present warming trend, which has been accelerating at a rate that is unprecedented for millennia (Global Climate Change- Vital Signs On the Planet –NASA) (Fig. 12.2).

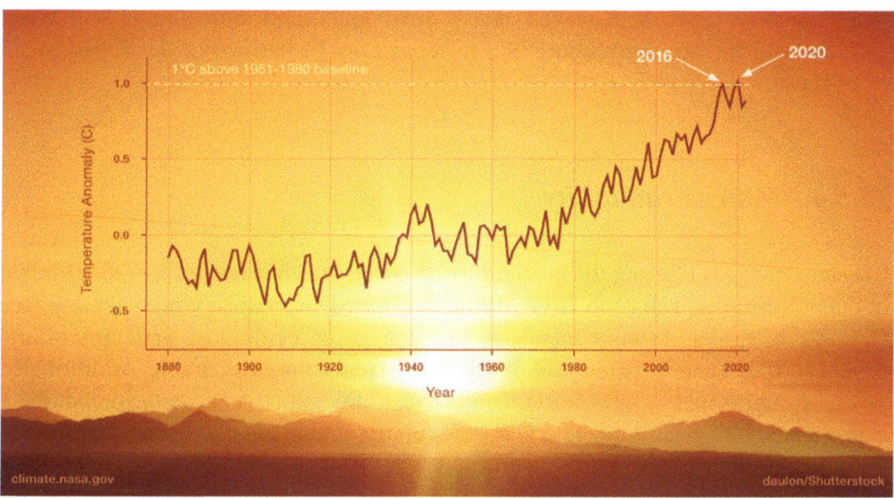

Fig. 12.2 This graph illustrates the change in global surface temperature relative to 1951–1980 average temperatures, with the year 2020 statistically tying with 2016 for hottest on record (NASA)

12.5.3.1 Causes of Global Warming

Anthropogenic causes including burning of fossil fuels, release of CFCs, use of vehicles, industrial development, deforestation, and natural causes like forest fires and volcanic eruption are responsible for global warming.

1. Disappearing glaciers and increase in sea level: Sea level rise and melting glaciers are two effects of global warming. The earth's temperature has increased dramatically. According to reports, Earth's average global temperature has risen by about 1 °C. As a result, there has been an increase in glacier melting, which has raised the sea level. The consequences for coastal areas might be catastrophic.
2. Climate change: Climate conditions have changed as a result of global warming. There are both droughts and floods in certain regions. The cause of this climatic imbalance is global warming.
3. Habitat destruction: Numerous plants and animals lose their habitats as a result of a worldwide climate change. The creatures in this situation are forced to leave their native environment, and many of them even go extinct. This is yet another significant effect of climate change on biodiversity. Many plant and animal species may become extinct if environments like coral reefs and alpine meadows are damaged.
4. High mortality rates: The average death toll often rises due to an increase in floods, tsunamis, and other natural catastrophes. Additionally, such occurrences may lead to the spread of illnesses that may endanger human life.
5. Spread of diseases: Heat and humidity patterns alter as a result of global warming. This has caused the migration of disease-carrying mosquitoes. Newly

bothersome pests, heat waves, heavy downpours, and more floods will affect crops and forests. Agriculture and fisheries may be harmed or destroyed by all of these.

12.5.4   Water Availability

Climate change is altering the availability of water, which is becoming more limited in many locations. The problem is further worsened for places, which already face shortage of water. It also increases the danger of ecological and agricultural droughts, which can harm crops and make ecosystems more vulnerable. Additionally, severe droughts and deforestation result in desertification. As deserts spread, there is less space for agriculture. The fear of regularly not having enough water affects a lot of people nowadays.

12.5.5   Availability of Food

Numerous factors, such as climate change and an increase in the frequency of extreme weather events, are contributing to the growth in global hunger and poor nutrition. Fisheries, livestock, and crops might all perish or become less productive. Due to the ocean's rising acidity, marine resources that feed billions of people are under jeopardy. Heat stress can decrease the amount of water and grazing areas that are accessible, which can limit agricultural yield and affect animals.

12.5.6   Health

Millions of people's health condition is anticipated to change as a result of the predicted climatic changes, including an increase in sickness, injury, and fatalities from heat waves, floods, storms, fires, and droughts. Malnutrition, diarrheal illness, and malaria outbreaks will make certain places more susceptible to severe public health issues, and long-term harm from catastrophes to health systems may jeopardize development goals.

12.5.7   Industry, Settlement, and Society

Typically, those located in coastal regions and river flood plains, as well as those whose economy are strongly associated with climate-sensitive resources, are the most susceptible industries, settlements, and societies. The economic and societal consequences of extreme weather occurrences will rise if they become more severe or frequent.

12.5.8 *Impact on Agriculture*

One of the industry's most vulnerable to climate change is agriculture. The effects of climate change on agriculture are becoming more and more obvious as a result of rising global temperatures, unpredictable rainfall patterns, and an increase in the frequency of extreme weather events.

12.5.8.1 Impact of Climate Change on Crop Yield

Certain areas, where temperatures were previously too low for optimal growth, may see an increase in crop production as a result of rapid climate change. However, because of heat stress, reduced photosynthesis, and increased water stress, higher temperatures can also result in lower yields in many places. Crop yields can be impacted by changes in precipitation patterns also, with some areas facing more frequent and severe droughts, while others could see greater flooding. Due to water stress, reduced nutrient uptake, and increased disease and pests load, crop yield reduction would be inevitable.

12.5.8.2 Changing Weather Patterns

The changing weather patterns associated with climate change also have an impact on agriculture. According to the USDA (2018), the unpredictability of rainfall patterns and the rise in the frequency of extreme weather events can lead to soil erosion, a loss of soil fertility, and an increase in crop susceptibility to pests and diseases. Reduced crop yields and food poverty may follow as a result.

12.5.8.3 Agriculture Water Scarcity

In many regions of the world, climate change is also causing an increase in the scarcity of agricultural water. It is becoming more challenging for farmers to irrigate crops as a result of higher evaporation rates and altered rainfall patterns (FAO, 2020). This, in turn, is leading to reduced crop yields and food insecurity.

12.5.8.4 Animal Agriculture and Livestock Rearing

Animal agriculture and livestock are significantly impacted by climate change as well. Heat stress in livestock can result in decreased productivity, illness, and even mortality as temperatures rise. Reduced fodder availability for animals due to changes in rainfall patterns can also result in higher feed costs and lower productivity.

12.6 Adaptation and Mitigation of Climate Change

In the context of climate change, adaptation refers to the actions taken to lessen the negative effects of climate change, such as relocating coastal communities in order to deal with the rising sea level or switching to crops that can withstand higher temperatures. Mitigation is taking steps to reduce the greenhouse gas emissions that are the root cause of climate change, such as converting to renewable energy sources like solar, wind, or nuclear power instead of burning fossil fuel in thermal power plants. Some strategies to deal with the climate change are as follows:

1. *Energy conservation and efficiency policies*: India can achieve its development and climate change goals with the support of policies that are physically possible, ecologically sustainable and economically feasible. There is a ton of potential for action on this front in all important areas. Energy efficiency initiatives "have the potential to reduce global greenhouse gas emissions by about 5–20% compared to baseline scenarios," according to the Intergovernmental Panel on Climate Change (IPCC, 2014). For instance, public transport networks should be promptly expanded to prevent an overreliance on the usage of private cars and to reduce the rising emissions in this industry. The use of biofuels should be encouraged, provided that they are not produced at the expense of food crops. Certain clean coal technologies that meet the criteria for cost-effectiveness could be used in the power generation industry. Increasing energy efficiency in buildings, vehicles, and industrial processes reduces energy consumption and pollution.

2. *Renewable Sources of Energy*: Increased usage of renewable energy sources, such as wind, solar, and hydropower, can assist cut emissions from the generation of energy based on fossil fuels. The National Renewable Energy Laboratory (NREL) published a report in 2012 that stated that "renewable electricity generation from technologies that are commercially available today... could supply 80% of total U.S. electricity generation in 2050, reducing greenhouse gas emissions by 80% relative to 1990 levels" (NREL, 2012).

3. *Agricultural Measures*: Adapting land use and management practices can help to mitigate climate change by increasing carbon sinks and reducing emissions from forestry and agriculture. According to the IPCC (2014), "mitigation options in the agriculture, forestry, and other land use sector can contribute to climate change mitigation by removing CO_2 from the atmosphere or reducing GHG

emissions from the sector." Additionally, according to a research by the Food and Agriculture Organization of the United Nations (FAO), sustainable agriculture practices including integrated crop-livestock systems and conservation tillage could cut world greenhouse gas emissions by up to 30%.

4. *The enactment of various programs*: National disaster management program, National Water policy of 2002, risk financing, etc. can help in mitigation and adaptation of climate change.

   - *Disaster management*: The National Disaster Management program coordinates disaster relief efforts and offers payments to victims of weather-related calamities. Additionally, it aids in the transmission of information and the training of those responsible for disaster management.
   - *Water*: According to the National Water Policy of 2002, in order to increase the amount of utilizable water resources, both conventional and unconventional methods of water utilization, such as rooftop rainwater harvesting and interbasin transfers, artificial groundwater recharge, and desalination of brackish or sea water, should be used. Water harvesting projects are currently required in many states and in many towns.
   - *The Forest Conservation Act of 1980*, which sought to stop the clearing and degradation of forests through a strict, centralized control of the rights to use forestland and mandatory requirements of compensatory afforestation in case of any diversion of forestland for any non-forestry purpose, sped up the afforestation process.

5. The key to maintaining yield stability will be the development of new crop types with increased production potential and resistance to numerous stresses (drought, flood, salt). One of the goals of breeding efforts should be to improve the heat tolerance of critical crop germplasm. In the same way, it is crucial to build tolerance to a variety of abiotic stimuli as they manifest in nature.

6. Changes to the cropping calendar to take advantage of the wet period and prevent extreme weather events during the growing season may be included in adaptation methods to lessen the negative effects of increasing climatic variability, which is often observed in arid and semiarid tropics. Cropping systems might need to be modified to cultivate acceptable cultivars, increase cropping intensities, or diversify crops.

7. Pest occurrence and severity on important crops will be impacted by variations in rainfall and temperature. This is because the interaction between pests, weeds, and hosts may shift as a result of climate change. Among the possible adaptation strategies are the following: (1) breeding pest-resistant cultivars, (2) integrated pest management with a focus on biological control and alterations to cultural practices, (3) pest forecasting using modern tools like simulation modeling, (4) alternative production methods, and (5) identifying crops as well as risk-free locations (Fig. 12.3).

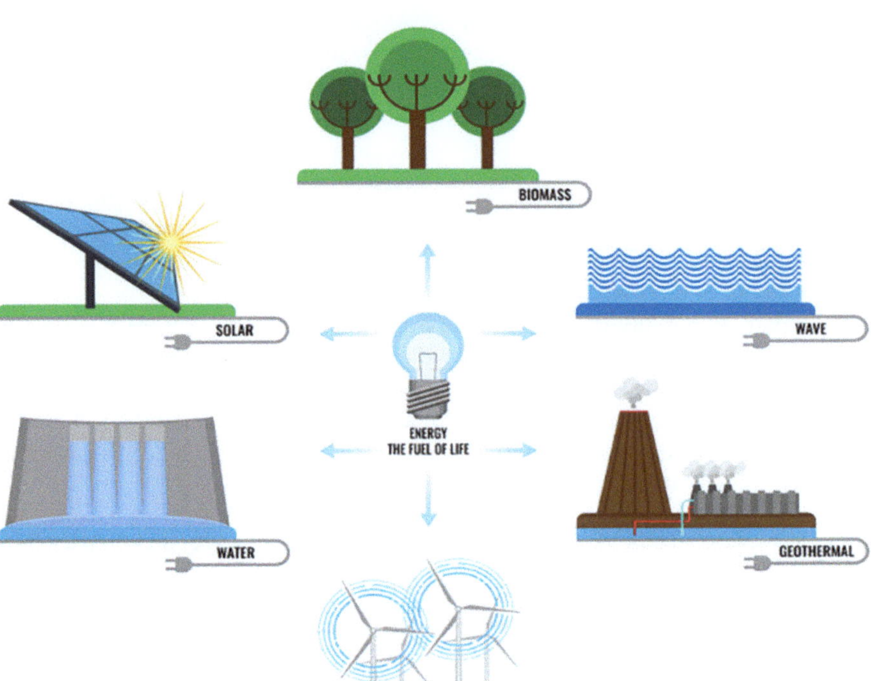

Fig. 12.3 Renewable energy sources (India Energy Portal)

MCQs

1. Climate change refers to:

   - Year to year fluctuations in the weather
   - Long-term changes in temperature and weather patterns
   - Short-term changes in temperature and weather patterns
   - None of the above

2. Climate change is also called:

   - Signal
   - Noise
   - Both 1 and 2
   - None of the above

3. Who developed the "greenhouse effects" theory?

- Tyndall
- J. B. Fourler
- Claude Pouillet
- Fischer

4. The average global temperature of Earth has reportedly risen by roughly

- 2 °C
- 0.4 °C
- 1 °C
- 5 °C

5. Which of the following is more destructive to ozone?

- Chlorine
- Bromine
- Nitrous oxide
- None of the above

6. Global warming leads to:

- Habitat destruction
- Spread of diseases
- Mortality
- All of the above

7. The ozone layer restricts

- Infrared rays
- Gamma rays
- Ultraviolet rays
- Visible light

8. Which of the following is not a greenhouse gas?

- Methane
- Nitrogen
- Water vapor
- Carbon dioxide

9. Which of the following causes is responsible for climate change?

- Change in solar output put
- Greenhouse gases
- Changes in the earth's orbit and rotation
- All of the above

10. The use of fossil fuels is responsible for the increase in the amount of which of the following gases?

- Carbon dioxide
- Nitrogen
- Ozone
- Argon

References

5th Assesment Report (AR5) – Intergovernmental Panel on Climate Change (IPCC) (2014) *Food security and food production systems* 95, 756, 810.

Chandrasekaran, B., Annadurai, K., & Somasundaram, E. (2018). *A textbook of agronomy.* 233, 234, 235, 236 - New Age International Private Limited

FAO. (2020). *The state of food and agriculture- overcoming water challenges in agriculture, 94.*

Mavi, H. S., & Tupper, G. J. (2004). *Agrometeorology – Principles and applications of climate studies in agriculture* (p. 264).

Mostafazadeh, R. (2018). *Re: Dufference between climate change and climate variation?* Retrieved from: https://www.researchgate.net/post/Dufference-between-climate-change-and-climate-variation/5a68a70bb0366d23404fff43/citation/download

National Renewable Energy Laboratory. (2012). *Renewable electricit futures study.* https://www.nrel.gov/docs/fy12osti/52409-1.pdf

USDA – United States Department of Agriculture. (2018) *Climate change, global food security, and the U.S. food system.* 20.

Index

MIX
Papier aus verantwortungsvollen Quellen
Paper from responsible sources
FSC® C105338

If you have any concerns about our products,
you can contact us on
ProductSafety@springernature.com

In case Publisher is established outside the EU,
the EU authorized representative is:
Springer Nature Customer Service Center GmbH
Europaplatz 3, 69115 Heidelberg, Germany

Printed by Libri Plureos GmbH
in Hamburg, Germany